22 GENERATIONS

22

GENERATIONS

INNOVATION, THE FERTILITY CRISIS, AND
THE APPROACH OF HUMAN EXTINCTION

WARREN HAYS, Ph.D.

DIADEMA PRESS   PORTLAND MAINE

Diadema Press
Portland, Maine
www.diademapress.com

ISBN: 978-0-9854182-7-4

Printed in the United States of America

to M

CONTENTS

INTRODUCTION

You hold in your hands a book that claims our extinction as a species is fairly imminent, due to – of all things – dwindling population. That might sound unlikely on the face of it, since one thing this world is not short on is people. In fact, if you're reading this in 2015 and you opened the book ten seconds ago, the human world has 24 more people in it than when you started reading. Our numbers are piling up fast, and they've been doing so for centuries now. It's probably hard for most readers to imagine the world running out of people in the foreseeable future. And yet, that's what's going to happen.

It was Thomas Malthus, an economist, who first warned that the global human population was set to explode beyond the world's ability to sustain it. That warning came in 1798, and has been echoed repeatedly ever since. When Malthus wrote that warning, the human population had already been growing exponentially without interruption for twenty straight generations.

Now, ten more generations have passed, and our population has continued growing exponentially right through them all.

Nonetheless, incredibly, the vast famines implied by Malthus (and explicitly predicted by others after him) have not come. We've seen appalling famines, certainly, but not enough to slow our long-term rate of increase. Malthus argued that our ability to improve farming technology couldn't keep up with our ability to produce children, so we must inevitably run out of food. As it turned out, he was wrong. Food calories available to the average person have actually increased worldwide in every generation from the time of Malthus until the time of this writing. Our population has grown by more than seven times since then, but our ability to produce food has grown even faster.

Why, then, should we believe that we're gradually going to stop reproducing over the coming generations, see our populations thin out, and finally allow our species to disappear? Given our extraordinary track record as a population-gaining species, what's going to change and make our immense sea of humanity dry up and vanish? What is already changing?

The answer to that set of questions follows a path that might seem odd at first, because it cuts through territory that has traditionally been claimed by a lot of separate disciplines. These include (mainly) anthropology, evolutionary ecology, economics, history, neurophysiology, sociology and demographics. If that sounds like it's going to be a meandering ride, I assure you it won't be. Our path will shortcut through these many fields precisely because it's going along the straightest possible route, ignoring boundaries and jumping fences along the way. We'll start in Chapter One with the room you're sitting in right now, and we will finish Chapter Ten with a clear and unimpeded view of the future extinction of our species.

Here is a thumbnail sketch of the argument this book is going to present:

Chapter One, *The Constructed World,* shows that modern human societies are packed full of innovations. This hoard of inventions and useful ideas has accumulated over a period of many hundred generations. Chapter Two, *The Accumulation,* explores the patterned fashion in which these innovations have built up, creating the civilizations we live in today. Chapter Three, *The*

Road to Happiness, examines this great trove of innovations from one especially important angle: as an essential basis of economic growth in the past, present and future. Chapter Four, *The Pleasure Principle*, re-examines civilization's hoard of innovations from a second, equally important angle: As efforts to reduce the feelings of dissatisfaction that arise from motivational circuitry in our brains.

Chapter Five, *On Having a Lot of Children*, hunts for the evolutionary purpose of the motivational circuits described in Chapter Four, and finds that they were biologically programmed to enhance reproduction among our ancestors. Chapter Six, *Getting Free from Natural Selection*, shows that innovations tend to short circuit our evolved neurological drives to reproduce by providing us with more efficient pathways to satisfaction. Chapter Seven, *Losing the Urge*, shows the effect of that short circuiting: fertility rates decline (in a highly predictable manner) as societies accumulate innovations and get rich.

With the basic issue established, Chapter Eight, *Natural Protection*, begins looking at some of the many things that might halt this trend toward infertility, particularly those that are natural, unmanipulated processes. We find that none of them will help. Chapter Nine, *Artificial Protection*, explores our several options as a species to boost our population by intentional actions, ranging from draconian laws to religious dictates to technological gimmicks. Remarkably, these all prove unable to block the progress of infertility in future generations. Chapter Ten, *The Extinction Spiral*, presents a model of future human population growth, showing that not only are we on a beeline toward extinction, but that we're going to get there surprisingly soon – about 22 generations from now.

So, that's the path this book will follow. If that line of reasoning sounds unlikely to you at first glance, then you're a particularly strong candidate to get something out of reading this book. Also: If you happen to be inclined toward the scholarly and/or technical, you'll find interesting material in the Notes and Appendices at the end of the book. In general, I've avoided putting graphs, equations and so forth in the main text, in order to keep the line of reasoning plain and clear for the casual reader. The mathematical materials can be found in the Appendices, while references to external sources and literature are in the Notes section, along with relevant digressions.

A few other preliminary comments, before we dive in:

1) The astute reader may already have noticed that I throw around the word 'generations' a lot, rather than speaking in terms of years. As mentioned above, the observations in this book come from several fields of thought, two of which are history and evolutionary biology. The word *years* invokes a very different feeling in each of those two fields.

When I think in terms of history, the phrase "six thousand years" conjures such gravity that I find it almost overwhelming. I find myself struggling even to imagine all those hundreds of empires, each passing through its entire history, each the home of a million forgotten stories and a thousand that are still written somewhere in the corpus of human literature... thousands of battles fought, millions of forgotten heroes. But when I put the same period of time into evolutionary terms, as 270 generations, it just sounds like a flash in the pan. Biologists are accustomed to thinking in units of hundreds of millions of years, not just a few thousand at a time and, after all, a colony of bacteria under your kitchen sink goes through 270 generations over any typical weekend. For me personally, as someone living embedded in the progress of history, the phrase: "Civilization began six thousand years ago," suggests such an awesome span of time that it makes me get a little weak in the knees. But change the phrase to "270 generations," and it sounds perfectly manageable.

So I'll be using generations as units of time quite a bit in this book. To keep things precise, I'll always use the word 'generation' to mean 22 years, when I'm referring to people. Admittedly, the reality of the situation is not nearly so tidy as that. There's a lot of discrepancy in human generation length from one time and place to another. On the other hand, the mean generation time of people is stabilized to some degree by biological constraints, such as the age of female reproductive maturity. In practice, anthropologists generally find that an estimate of 20 years per generation is a bit short and 25 a bit long, so 22 is actually a reasonable compromise.

One great thing about choosing to use this imperfect unit of measurement, the human generation, is that it allows us to directly compare what's happening in humans with what happens in other species, evolutionarily. Generations, not years, are usually the appropriate units for looking at rates of mutation and natural selection.

2) This book will simply ignore supernatural explanations. We'll act under the assumption that there are efficient natural explanations for pretty much everything, even if people haven't uncovered all the details yet. The most important categories of supernatural explanation that we'll be ignoring are those that come from religion and Platonic ideology. The first of these two is probably a familiar exception to every reader, including those with religious beliefs. Almost all books written by intellectuals, scholars and scientists in the past century or so have gently (or roughly) pushed religion out of the way as a preliminary to serious discussion. I'm not interested in denying the validity of anyone's faith, but we'll limit ourselves to natural explanations in this book.

The second category of supernatural explanation that we'll be excluding may be a little less familiar. Plato proposed that ideas exist in a world separate from that of physical reality, and that they are in some sense more 'real' than the reality we experience through our senses. This gives ideas (in his view) an almost god-like power to cause patterns and actions to occur in the physical world. But we're going to simply ignore any notion that pure, ideal forms can loom above the material plane, free from it, and yet act as templates that affect how things occur down here in our home world. After all, if they can do that, then these pure ideas are certainly above nature (that is to say, they are supernatural), so this Platonic viewpoint is in many ways indistinguishable from a religion or superstition.

Some ideals of the Platonic sort have been used so widely that they bear specific mention here. Plato's own chief ideal, "the Good," may seem a little outdated and naive to most of us jaded moderns, but probably a lot of us still like to think that there's some sort of universal tendency for Good to win out over Evil. I'm not going to deny it, any more than I'm going to deny the validity of any specific religion. This book will simply move forward under a different set of assumptions: those of efficient natural explanation.

Similarly, we'll ignore the idea of Fate, or (if you prefer) Destiny. A third specific example, from Nietzsche, is the Will to Power. Of course, there's room to argue that Nietzsche was just describing a natural tendency, not proposing a Platonic, supernatural, causal force. But I don't think you can *read* Nietzsche and still believe that. Anyway, *pax*, we don't need to argue the matter one way or another. To the degree that the 'will to power' is merely a natural phenomenon, it will fit right in with this book's point of view.

3) The word "evolution" will be used in two distinct ways in this book. One is the biological sense, meaning a directional change in gene frequency in a population. If that last phrase means nothing to you, don't worry – we'll have a closer look at the concept in Chapter Five. The other use of the word evolution is in the phrase 'cultural evolution', which can mean lots of different things. Under most definitions, 'cultural evolution' has nothing to do with evolution in the biological sense. We'll see one or two exceptions in passing, but they won't be crucial to this book's point.

Although I'm an evolutionary biologist by training, my project here is not to claim that civilization is a biologically evolved structure. In fact, I'm committed to the view that it is *not* one. I think it's pretty clear that we humans built the whole thing, piece by piece, and that we planned each piece in advance. Advanced planning is exactly what natural selection doesn't provide. Natural selection performs its miracles in the utter absence of intelligent design. On the other hand, I think that basically everything in civilized societies *is* the result of intelligent design, and the intelligence is ours.

4) I'll bet there are a few readers who are shifting uncomfortably in their chairs, now that I've said the words "civilization" and "biologically evolved" in the same sentence. Let me put you at rest: I'm not building up

steam to propose some kind of eugenics program. Far from it.

If, on the other hand, you're a reader who is not sure why it might be scary to see an evolutionary biologist writing a book on culture and sociology, here's a quick summary of the problem. A few biologists from the late 19th to the mid 20th centuries (most notably Julian Huxley), held forth the view that the poorest people in civilized nations were genetically inferior and should be weeded out. This view came to be called 'social Darwinism', and efforts to enforce it were known as 'eugenics'. The US was the great pioneer of eugenics. Forced sterilization programs operated under legal mandate in the US from 1907 until at least the 1970s. Not everyone supported the idea, so the matter was brought before the Supreme Court in 1927. The court declared that forced sterilization of the mentally incompetent was fully constitutional.

The policy spread from the US to many other nations, including (most notoriously) Germany in the 1930s. Non-biologists who became enthusiasts of these schemes included the politicians Theodore Roosevelt, Herbert Hoover, Winston Churchill, and Adolf Hitler. Other notable proponents of eugenics (real geniuses in their fields, but sadly ignorant of evolutionary biology) included economist John Maynard Keynes, writers H. G. Wells and George Bernard Shaw, and the inventor of the transistor, William Shockley. The policies of eugenics reflected deep misconceptions of how Darwinian natural selection actually works, and anyone who supported them didn't understand evolutionary biology.

Anyway, as I said, the reader needn't worry that this book's observations can be misconstrued as social Darwinism. The problem addressed in this book can't possibly be blamed upon the poor. If anyone is to be held responsible, it will have to be those of us who are fully enfranchised in our society, those of us who are supposed to be in charge around here: the intelligentsia, the leaders of government, the rich. But ultimately, there will be no real blame to lay on anyone, because we're going to find that the trend is a direct consequence of human nature.

5) One last comment before we begin. As you read this book, at times you might feel that my attitude seems a little glib for a person who is

predicting human extinction. It might seem, starting out here, that a gloomier, perhaps even funereal tone of prose would suit better. But there's another surprise that this book has in store for the reader. It will turn out that the events predicted here won't come upon us as a disaster, but rather as an immense act of collective intent – the exercise of our empowered preference. All of us are going to do exactly what we want to do.

We as a species are going to proceed toward extinction with our eyes wide open, smiles on our faces, and no one forcing us to do anything. We will *want* things to go along just so. In fact, the desire is already measurably growing inside us, and, as we shall see, we have already begun to act accordingly. It's strange to say, but this will turn out to be a book about happy, contented people in the not-so-far future, doing exactly what they feel like doing.

Part One

GETTING OUR WAY

1

THE CONSTRUCTED WORLD

You are sitting in a room. You are surrounded by materials, forms... there's some wood, some plaster, some glass, cloth, metal. Almost everything around you shows the clear signs of human craft. The glass was once sand, the metal was ore, the planks were trees and the cloth – well, who knows what the cloth started out as? The hair of a sheep, maybe, or a fibrous mass extruded in the fruiting bodies of cotton plants. More likely, nowadays, it was pumped out of the ground as fossil hydrocarbons, distilled in a refinery column, then polymerized into fibers of polyester or some such thing.

Remarkable, isn't it, how utterly manmade is your immediate environment? Unless you are one of the extremely few readers who happen to be enjoying this book under a spreading tree by a brook somewhere, you are quite likely to be unable to spot even one thing – not even *one* – that's in its natural state, unless you look out a window. If you're in the city, even that may not do it. We, modern humanity, have truly built ourselves a world. Now, we live in it... and as for nature, well, nature is out there somewhere, even if we tend to forget about it most of the time. The notion that we as a species used to actually *live* out there is almost beyond belief. Paleolithic 'cavemen' seem freakish to us, and it's easy to see why. The notion that we humans were actually like that, once upon a time, is hard to swallow.

While we're in this reflective mood, let's take a quick look at one of the familiar objects that surround us here in this room. Where on earth did all this stuff come from? Somewhere near your chair there's an electric lamp. The bulb may very well be fluorescent or LED-powered, but even today it's probably the old filament type. That's an easy starting point, because we all know where that invention came from: Thomas Edison. We may even know the famous date when the light bulb was born, namely October 21, 1879. So we can say this much, right off the bat: from the origin of our species, six thousand generations ago, right up through October 20, 1879, there was no such thing as a light bulb. The day after that, light bulbs existed. Today, five generations later, I can say with pretty good confidence that if you're indoors, there's a light bulb somewhere nearby.

Thomas Edison didn't build the actual light bulb beside your chair, obviously, but nonetheless there is a sense in which his contribution, his innovation of that technology, is embedded within it. He "left his mark upon the world," as we say, and that mark is right there in your home. As you look around yourself again and notice how few things in the room are in their original, natural state, you're bound to realize how surprising it is that *each* of these objects bears the mark of some brilliant innovator in the past. In fact, most of those objects must be stamped by many such innovators, a lot of whose names have been lost to history.

Have a look at the book in your hands... not as a source of readable information, but in its physical form: a chunk of bound and printed sheets of paper. If you're reading these words in digital format, don't even get me started on the innovations you're holding before your eyes! But if you've got a book made of paper, then you might call to mind the name of Johannes Gutenberg, who invented the movable-type printing press in 1439. True, there's not a chance that the book in your hands was printed on such a press, not in this digital age, but in a certain sense Gutenberg's fingerprints are still recognizable.

So, what other innovations did it take to make this book? Well, there's the paper, which someone invented in China about 1800 years ago. Someone later invented the wood-pulping process that led to lignin-based (rather than rag) papers. Someone invented calendering to flatten the paper surface. Then there's smudge-proof ink, and digital reproduction, and hot-glue binding – and the various things that preceded that form of binding such as saddle-stitching and signature stapling.

But all of that is just the tip of the iceberg. Someone invented the symbol system we call the Latin alphabet, deriving it from the Greek alphabet about 2300 years ago, as a system for rendering the phoneme elements of speech into visual code. And here's a big one: someone innovated *each word*. That means that each and every unique word in this whole book is a separate and independent innovation thought up by some clever person in the far past. That one new word was very possibly the smartest and most lasting thing that particular person ever did. And then, of course, there's our system of grammar, our system of syntax, our huge collection of idioms, our literary conventions of irony and rhetoric and metaphor... you begin to get the idea.

So the book in your hands contains the palpable marks left by literally thousands of specific, pinpoint innovations, each one of which is the product of somebody's bright idea at some particular moment in the past. Prior to that historical moment, that particular innovation simply did not exist. This book embeds them all... which makes it kind of unfair, I suppose, that my name is the one that ended up on the cover.

Next time you're walking through a modern city, consider how outrageous, how awe-inspiring, is the fact that you can hardly see anything, anywhere around you, that doesn't have that same astounding quality of rich embodiment of innovations from the past. Not a skyscraper, not a fire hydrant, not a scrap of pizza wrapper in the gutter: each is bursting with the innovative brilliance of individual people from the past – and even some who are still living. Only the tiniest fraction of that body of innovations was inherited from our Paleolithic ancestors and their natural environment.

It's impossible to count how many innovations went into building up our synthetic urban environments and the cultures that surround them, but here's one indicator: Not long before this writing, the US issued its nine-millionth patent. And that still only accounts for a very small fraction of the innovations around us, because most innovations are not patented.

Figure 1. Humans are an innovative species.

We are surrounded by valuable things, and a lot of their value is due to innovations that have built up over many generations. Each innovation can be traced back to some individual or small group, who first thought it up. That's a fundamental connection between the minds of individual people and the economies of nations. The picture here is a classic press photo of Nikola Tesla in his lab, taken by Dickenson Alley.

Humans are not the only source of innovation in the world. Consider a spider web, a work of apparent genius that is created every day by exceptionally stupid creatures. The brain of a typical spider weighs about as much as a single speck of milled flour. When you think about it, it's kind of surprising that such a creature has enough neurological wherewithal to swallow its own food, much less build something. And yet this shockingly stupid creature has just enough memory and control circuitry built into that tiny brain to allow it to carry out the series of operations that stick its silk together into a functional web, often creating a beautiful symmetry.

If we journeyed back 400 million years, we'd find ourselves living in a world where there was no such thing as a spider web, nor had there ever been anything of the kind. Somewhere between those ancient days and the times we live in now, webs appeared as an evolutionary innovation. The process wasn't abrupt; it was a gradual building up of small innovations that kept improving the functionality of webs, until today's webs came to exist in all their amazing forms.

Spiders didn't *think up* the idea of spider webs. The notion that some spider genius in the distant past had a bold vision in its tiny brain and got to work tinkering on the first web, that's clearly ridiculous. Rather, a long process of natural selection operating over millions of generations distilled a neat kit of behavioral-instinct genes from among the randomly mutated bits of DNA code found in the ancestral spiders, giving some spiders a competitive advantage because they could catch more food. It was a process of blind tinkering, with no more intelligence behind it than a rock rolling downhill. Nonetheless, spider webs eventually came to exist, just as surely as did light bulbs.

Let's compare the innovation of the first spider web with the innovation of the first light bulb. There was no biological evolution involved in the first appearance of light bulbs – no process of natural selection among randomly mutated DNA. Edison had the idea, which led him to come up with a plan, after which he executed purposeful actions until that plan bore fruit. Spiders don't work that way, but we do.

Activities that follow a plan toward a pre-determined goal, like Edison perspiring over his light bulbs, are called *teleological* activities – basically just a fancy word for 'purposeful', since teleology literally means "the study of purpose." Planned actions are always teleological, because we only carry out a plan when we have some purpose in mind. Human innovations are not always teleological (sometimes people just stumble across something useful), but more often than not, innovations do occur in human societies as a result of purposeful activity that arose from some sort of planning.

Teleological explanations of innovations can run into problems, however, because they can easily get out of hand. If I say if that the *idea* of the light bulb existed before the actual light bulb existed, that's an easy opinion for most people to agree with. Edison and his colleagues wouldn't have done their experimental work if they hadn't known what they were trying to build. But we can get into trouble if we try to apply that same sort of teleological explanation on the grand scale.

If I claim that the idea of *civilization* was around before civilization existed, then that's going to lead me into the realm of supernatural explanations. No one believes that some forgotten genius cooked up the whole idea of civilization and labored mightily until he built the first one out of nothing. Rather, when someone claims that the *idea* of civilization was there before the real thing appeared, they either mean that the idea existed in the mind of some deity, or that it was written into the invisible fabric of fate (maybe in Plato's "world of forms"). Because these opinions attribute the origin of civilization to supernatural causes, teleological explanations of cultural evolution are unpopular among modern anthropologists and sociologists.

Things weren't always like that. Early theories of the origin of civilization tended to be very teleological. The first such theory seems to have been put forward by Anne-Robert-Jacques Turgot in a lecture at the Sorbonne University in Paris in 1750. This was at the height of the Enlightenment, with the Industrial Revolution just warming up, and Turgot's lecture is believed to have been the first time that any person anywhere put forth the idea of 'progress', in the sense of cultures becoming more developed over time. Nowadays, for most people, the word 'progress' conjures up ideas of social and economic improvement of the human condition, but for Turgot, cultural advancement was largely religious. He was concerned with our rise from the status of lowly heathens toward the lofty pinnacles of Christian perfection.

Turgot told his Sorbonne audience that humanity can be traced back to the single family of Noah, which survived the flood described in the Book of Genesis. That family split into tribes, and, subsisting in barren deserts among wild beasts, they were "plunged into barbarism." Turgot pointed to the tribal cultures of Native Americans as evidence. Civilizations slowly emerged from those barbarian tribes by a process of cultural advancement punctuated by five key innovations, in the following order: agriculture, politics, writing, philosophy, Christianity.

Turgot's story is teleological because what he was really doing was explaining to his own culture (18th century France) what its purpose was. The story of history that he unfolded was merely the backdrop to a pep talk. He finished by telling his audience that they, as top intellectuals, had been handed the torch of cultural advancement, and they were now duty-bound to pass their special French brand of intellect and Christian religious fervor to the world.

Similar teleological feelings of cultural purpose are common in nations with imperial ambitions, where people feel that their upcoming conquest of other nations is simply written in the stars. This attitude was prevalent among European nations during their centuries of global colonization (late 15th to early 20th centuries), and was made explicit in the 19th century American catch-phrase, 'manifest destiny'.

So the word 'teleology' tends to make scientists bristle, but only because the idea is ripe for abuse, and has a long history of getting in the way of clear understanding. But at the same time, everyone is happy to admit that on the small, individual scale, the human process of innovation *is* teleological. In fact, small-scale teleological innovation is the big difference between humans and other species. As we saw in the case of spider webs, innovation is certainly not unique to the human species, but *teleological* innovation is extraordinarily rare in other species, while it's very common among humans.

True, a few cases of non-genetic, intelligence-based innovation have been observed among semi-wild primates. Food washing was invented among Japanese macaques by a young female named Imo in 1953, as surely as Edison invented the light bulb. That suggests that some other techniques among primates may be the result of innovations passed down from specific inventors in previous generations. Termite picks as used by chimpanzees are a likely example. But the total number of such cases in all nonhuman species is certainly tiny compared with the number of innovations that led to, say, the space shuttle.

If other organisms could routinely plan their innovations, rather than usually achieving them through a mindless intergenerational tinkering upon the genetic variation in their instinctive behaviors, then they would be likely to select those innovations which alleviate immediate wants. A squirrel in winter, if it possessed the capacity of imagination needed to achieve purposeful innovation, would far prefer to build itself an A-frame cabin with a Franklin stove to get through the icy months, rather than curling up in a ball of leaves. We humans have an ability to see the impediments that are keeping us from getting what we want, and that lets us come up with our plans, apply our disciplined intelligence, and innovate to improve our lives.

From one point of view, then, our modern civilizations really *are* the result of purposeful intent after all... it's just that the purposefulness was applied to one little problem at a time. The origin of civilization had no grand teleology, true, but it was teleological nonetheless. Civilization is a huge heap of tiny bits of planned, purposeful innovation, assembled bit by bit. We may not have had any great vision before we started building civilization for ourselves, but gradually we've piled up quite an edifice by solving one problem at a time.

We've seen that a lot of familiar objects, such as light bulbs and books, are the result of long series of innovations that occurred during historical time. In general, manufactured material goods differ from the raw, natural materials that they're made from due to a series of modifications, each of which originated in the past as an innovation. In this sense, the manufacturing process has embedded the innovations into the finished light bulb or book or whatever.

I'll be using the term **embedded innovation** to describe this implied presence of a past innovation in any artificial object or concept or activity. At first, this may seem an unnecessary bit of jargon, but I have a strong motive for introducing it, so bear with me for a moment.

Let's start with a concrete example. A wooden chair is an artificial good, and a tree trunk is a naturally occurring commodity, but both are made of the same substance. And yet, from the viewpoints of people, cultures and economies, the chair is something more than just a piece of tree trunk. The chair is worth more than the material it's made from, precisely because that material has been modified – in fact, modified over and over. The original piece of tree trunk was converted into rectangular, treated pieces of lumber in a sawmill, and then the lumber was elaborately worked in a woodshop, piece by piece, into a chair. From beginning to end, hundreds of steps were performed between the forest and the furniture store. Each of those steps, each modification, could (in principle) be traced back to a specific historical origin: the day when someone first figured out how to do that particular job in that particular way.

Another useful way to express the difference between the piece of tree trunk and the chair is to say that productive labor changed the raw wood into furniture. I agree with that view entirely, but I'd like to go one step further. The productive labor wasn't just a certain number of paid hours of random exercise by workmen; rather, every step was a precise, purposeful, trained activity. The labor of building the chair consisted of going through a list of modifications to the wood, each of which used a technique that's part of our cultural heritage, and each of which first appeared as an innovation on some specific occasion in the past. The labor of making the chair was the process of embedding those innovations into the wood, and thus converting it into something more valuable.

The concept of 'embedded innovations' is not the only way to look at these matters, and for many purposes it would be easier and more straightforward to say that the difference in value between a chair and its raw material is a matter of labor, or capital investment, or other things. But for the sake of this book's project, the idea of embedded innovations is going to prove very useful because it will eventually connect the *source* of the innovations (problem-solving individuals who were attempting to make their world a better place), and the *outcome* of embedding innovations into objects and labor practices and so forth: namely, increased economic value. The concept of embedded innovations will keep our eyes on the link between issues of personal satisfaction and the broader issues of socioeconomics. Necessity is the mother of invention... and invention is the rocket fuel of economies.

The chair example is straightforward, but there are also more subtle ways in which innovations can become embedded in things and practices. For one thing, manufactured goods aren't the only objects that embed innovations, even though they're the obvious examples you see when you look around your room. Another important type of example is raw materials, or commodities. Almost all of the so-called 'raw' materials used by industry and consumers are not naturally occurring, but have been modified by techniques that first appeared as innovations, adding to their value. A pile of gravel is nothing but raw material from the viewpoint of an architect or a road builder, but in fact it's already got a remarkable number of innovations embedded in it. It was quarried up out of a rock seam with diesel equipment, then processed through an immense crusher, then sorted by size, and delivered. There are a lot of very profitable patents from past years lurking in a pile of gravel, and (similar to the case of the chair) those embedded innovations actually make up *more* of the gravel's value than the underground rock that was the natural source material.

We can go even further. Even the truly raw, natural source materials that are still sitting out in nature, untouched by humankind, can have a value embedded in them by innovation. That value can become embedded, as if by magic, even before anyone discovers a hidden deposit of natural materials. The reason for this is that the historical appearance of an innovation can change the value of naturally occurring materials drastically. For example, until a given society has discovered how to smelt iron, iron ore will be regarded as just a bunch of rocks, but afterward it will become a valuable material in high demand. Similarly, once the internal combustion engine was invented in the late 19th century, every fossil oil deposit (most of them undiscovered at that time) suddenly became 'black gold'. The value of the innovation was not only embedded in the early Benzes and Fords, but also in the hidden seas of oil under Texas and California.

Another place where innovations can turn up embedded in the civilized world around us without being particularly noticeable is among intellectual properties. The term 'intellectual property' refers to valuable assets that have no definite physical form. For example, a proposal for a new shopping center may be printed in a brochure, or kept in a digital file, or simply communicated in a stream of hot air coming out of someone's mouth. Whatever physical form embodies it, the proposal has its own value.

A given, particular proposal for some shopping center is no more likely to be original and innovative than, say, a particular chair. Nonetheless, like the chair, the proposal has huge numbers of past innovations effectively embedded inside it. In addition to all the conventions of language, from the alphabet to the rhetorical styling, a proposal for a shopping center is likely to embed a lot of innovations from other fields, such as economics – assuming the writer hopes to coax investors to stump up some money.

So plenty of examples of embedded innovation can be found in intellectual properties such as patents, manuscripts, computer programs, and so forth. But probably even more innovations are embedded in abstract items of our culture that *can't* be called intellectual properties because there's no way to put a price tag on them. Many abstract items are valuable to a society but can't be regarded as assets because, like the air we breathe, there's no way to assign them a monetary value. An example would be a new turn of phrase, which emerges from somebody's mouth in casual conversation, makes the rounds, and eventually spreads throughout the culture. After a while, everyone may know the phrase and use it as a casual idiom whenever they find it useful, even though no one knows where it originally came from. Other examples would be the visual conventions that come and go over the years in the graphic arts, and the ever-changing speech mannerisms of youth culture, and the fussily particular table manners of the wealthy. In all such cases, large numbers of innovations are embedded in abstract forms that add structure to our culture, but can't be priced as assets.

So, we've seen that innovations are embedded in objects, materials and abstractions. Now we come to the big one: labor practices. For one thing, all of the embedded innovations found in a manufactured object are the result of particular labor practices which themselves embed the same innovation. A wooden chair has legs that have been joined to the seat in a particular way, and that method of wood-joining arose on some specific day in the past as an innovation. Another way to express the same thing would be to say that that particular woodworking technique arose as a labor practice on that day. The innovation is actually embedded in both: the labor skill and the product created by that skill.

Furthermore, there are many embedded innovations in modern labor practices that are invisible in the final product. They increase the efficiency of making something without necessarily altering the item that's being made. These embedded labor innovations were the key to the Industrial Revolution, and since then have transformed the wealthy nations into industrial societies. The first major examples appeared in England in the 18th century, in the form of innovations that accelerated the processes for spinning wool or cotton fibers into thread, and for weaving the thread into cloth. Both of those processes had been done by hand for centuries, using spinning wheels and hand looms that were usually kept in individual homes. Automated weaving began with the innovation of the flying shuttle in 1733, and automated roller spinning began in 1742. By the end of the century, both processes were powered by steam, and England was manufacturing cloth in the world's first big, automated factories.

Conceivably, we might be able to hold two pieces of cloth in our hands, both made in 1800, one made entirely by hand and the other made entirely on factory machines, and we might find no differences that meet the eye. This dramatizes the fact that the automation process didn't actually embed any innovations into the *product*, but rather embedded them into the labor practices that made the product. The Industrial Revolution didn't necessarily improve the average quality of cloth, but it drove England's economy to previously unimagined levels of wealth.

The assembly line is another familiar example of an innovation that is embedded almost entirely in labor practice, rather than in the finished products of that practice. We can also find a whole set of such examples embedded in farming practices, including the use of large combine tractor systems, synthetic fixed-nitrogen fertilizer, organophosphate pesticides, genetically modified crop strains, etc. The food found in grocery stores may look more or less the same, but the number of man-hours required to grow it has plummeted since the Industrial Revolution, and modern farmers use a very different skill set than farmers of the 18th century.

The automation of factories and farms has embedded modern labor practices with countless innovations, some of which are not likely to *seem* like improvements from the viewpoint of individual workers. Repetitive labor on an assembly line is famously less appealing than old-fashioned craftwork in a shop, even if it gets the same job done faster and more cheaply. However, this same process of automation and refinement of production techniques also creates a demand for *some* workers whose skill sets embed huge bodies of innovations. These include engineers, technicians, computer programmers, factory managers, and so forth. Similar roles for highly skilled laborers have also arisen in the service sector of wealthy industrial societies, and include physicians, lawyers and scientists, among others. In all of these cases, 'skilled' labor can be described as labor that contains a lot of embedded innovations.

The difference between 'skilled' and 'unskilled' labor carries a market price, and this price is strongly correlated with the amount of training needed to acquire the skill. The amount of time it takes to acquire that training is probably a good indicator of how many separate innovations have become embedded in the skill set over historical time. Surgeons, for example, are manual laborers who must complete 25 years of school and apprenticeship before they are licensed to practice. When they get to that point, their labor is valued at 16 times as much as the work of an 'unskilled' laborer. That difference in labor value is largely due to the innovations that have been embedded over the course of history into the surgeon's skill set. It's important to remember, however, that there really is no such thing as unskilled labor. There is no work environment in which a worker can be productive without exercising *some* acquired skills, which is to say, some embedded labor innovations.

One last note on the nature of embedded innovations: Some innovations have been replaced by subsequent, improved innovations over and over again. The result can be a single embedded innovation of extremely high quality, in the sense that it adds a great amount of value to an object or practice. Occasionally, these high quality innovations may also appear as sudden strokes of genius, but much more often they are the result of a long historical process. The earlier, obsolete versions of the innovation may not be evident in the modern object, as we turn it over in our hands and look at it, but their value is nonetheless embedded there. This progressive accumulation of innovations will be the topic of the next chapter.

2

THE ACCUMULATION

We've seen that we're up to our ears in embedded innovations. They didn't accumulate willy-nilly, or at least not all of them. Some of them were laid down very early in our cultural progress and served as foundation stones for others. More than that, in many cases the order of appearance doesn't seem to have been arbitrary: particular innovations (like agriculture) really *had* to appear early on, or most of the other bright ideas (like castles and airplanes) would never have appeared at all.

Exactly *why* that should be true is the basis of some of the most contested intellectual speculation of the past two hundred years. Luckily, we can count on wide agreement if we just stick to this very general statement: There are certain innovations that had to appear early in the accumulation, or most of the other innovations wouldn't have shown up later on. There's also broad agreement on the identity of two of these key innovations: agriculture and cities. Most of the innovations around us today couldn't have arisen before the appearance of agriculture (first) and cities (second).

At some point during that progressive accumulation of innovations, a society comes to a level of complex organization where we call it a 'civilization.' Unfortunately, there's not much agreement on how we should define that word. A lot of dictionaries define civilization as something along the lines of "a developed or advanced state of human society," but there are probably no anthropologists anywhere who would accept a definition of that sort. Words like 'advanced' are not only vague, they're also dripping with teleology. They seem to suggest we're all marching toward some sort of goal, and have been from the start. The most straight-forward way to define the word 'civilized' is by its literal meaning, which is: "characterized by cities." Perhaps half of the world's anthropologists would be content with that definition, and that's about as much as we can hope for. For the purposes of this book, and for the sake of simplicity, I'm going to use the word 'civilized' to describe any culture that contains cities, and non-civilized to describe those that don't.

Looking back over time, we see that cities can first appear in a region either by arising there spontaneously or through influence or colonization by other civilizations from abroad. If cities arise spontaneously, we call the phenomenon an original civilization. There have only been seven original civilizations, and each has appeared in a place where agriculture was already well established. The seven locations were Mesopotamia, the Indus River, Egypt, China, the Andes, Central America, and the southeastern portion of North America (see Figure 2). Two of those spontaneously collapsed, namely the Harappa culture of the Indus River valley, and the Mississippian culture of North America, and in both cases the region reverted to a non-civilized condition. The other five have remained active centers of city life continuously to the current day, though the ones in the Andes and Central America lost almost all of their original cultural attributes when they were overrun by Spain in the 16th century. Spain's civilization, incidentally, came out of Mesopotamia (via the Assyrians, Phoenicians, Greeks, and Romans, in that order).

Only a few things were indisputably present in every one of the seven cases *before* the first city appeared, and two are worth special mention:

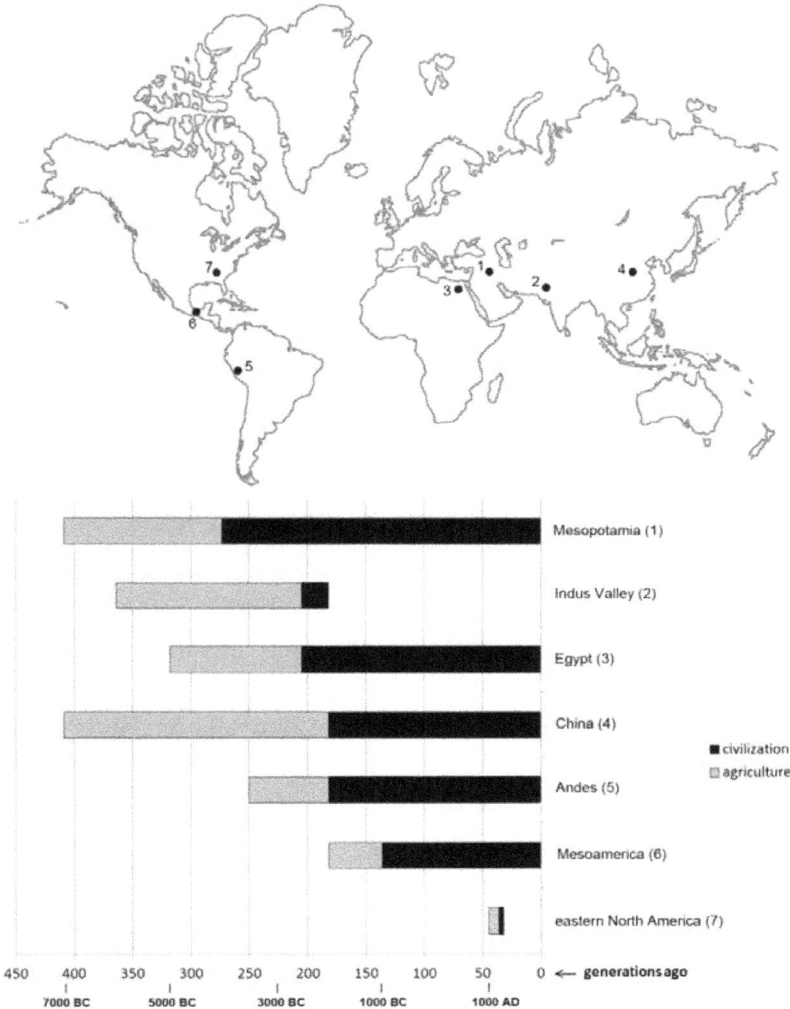

Figure 2. The seven known original civilizations.

Notice that two of them have gone extinct. Others, specifically the Andean and Mesoamerican civilizations, have been so drastically altered by contact with outside cultures that they have been effectively annihilated by absorption. See the Notes for details, discussion and references.

agriculture and tax. It's clear that these three early innovations – agriculture, tax and cities – played powerful roles in the early formation of our modern hoard of innovations, but we're sadly short on evidence to clarify how those three things came to exist, or exactly how they influenced the development of civilization. The ratio of theories to data in this field is absurdly top-heavy. Archaeologists have done a lot of digging for evidence, but that sort of work always advances slowly. Not only are the key dig sites hard to find, they tend to be very old, and the settlements were usually built out of non-durable materials like wood and thatch, and (being pre-civilized) they were never very densely inhabited in the first place. We may never have enough direct evidence to answer all of our questions with complete confidence.

When scientists can't find enough direct evidence to answer a question, they often try to fill in the gaps by making comparisons among the things they *can* see. In this case, that involves using the small, non-civilized cultures that still exist in a few remote places, hoping that we can compare them with the long-gone Neolithic cultures that led to the original civilizations. There are a lot of ways to go wrong in making such comparisons, but if we don't try, then we're stuck reconstructing the entire origin of civilization from the data dug up at archaeological sites... and those data are mighty sparse.

Of course, one expeditious thing we might do is just randomly select some hunter-gatherer society that's still living in today's rainforests of Indonesia or Brazil or Congo, and use it as an example, hoping that pre-agricultural Mesopotamians might have been roughly the same. But there are good reasons to proceed with more caution than that. For one thing, hundreds of modern hunter-gatherer cultures have been studied already, and the variety of traits they show is mind-boggling. If they're not all the same as each other, then surely none of them is *exactly* like the lost pre-agricultural societies we want to know about. As if that weren't bad enough, almost all existing cultures have had some degree of contact with civilizations, and these contacts have often altered them substantially and subtly. Europeans have been wandering all over the world for five hundred years.

Fortunately, there is *some* archaeological evidence from every region where agriculture first appeared, and also from each of the seven original civilizations. By combining that direct evidence with a cautious comparative approach, anthropology has managed to establish some aspects of the story of civilization's origins pretty firmly. Let's consider the elements of that story which can be asserted with the most confidence.

The original human social units were clan groups called 'bands' that lived as hunters and gatherers. These extended-family societies almost always contained fewer than a hundred people. The band was the basic human social structure from the origin of our species, 6000 generations ago, until agriculture appeared. Today, band sociality may or may not be extinct, depending on who you ask, but at least until very recent times there were still a handful of examples. These included the Bushmen and the Mbuti pygmies in Africa, and some aboriginal Australian groups.

Certainly, one thing that's intriguing about band sociality is that it was the only form of human social structure for 90% of our species's history. A second thing that's intriguing is that bands – that is to say, kin groups that live and work together for mutual benefit – are the only form of human social group that has close parallels in other species. In fact, as we'll see in Chapter Five, there are quite a few species that live together in kin groups that are about the size of human bands, and kin-based sociality is very well understood by animal ecologists. All the other forms of human social system, such as modern industrial nations, are hard (if not impossible) to explain in terms that would be comprehensible to a biologist.

A third thing to notice about the band cultures that preceded agriculture is that they were already accumulating innovations, and had pretty good hoards of them, even before they started tilling the soil. This is true despite the fact that very little accumulation or concentration of material and labor assets takes place in band societies. For the most part, food is gathered and consumed without mass storage, villages tend to remain unfortified, slavery is nonexistent, and individuals don't pay tax to their leaders. Nonetheless, a lot of innovations did accumulate in each band culture prior to agriculture. Examples include projectile weapons such as slings, bows and blowguns; tang systems for attaching arrowheads and spearheads to wooden shafts; and fiber-making and weaving techniques. A few pre-agricultural innovations, such as clay-based ceramics and fish hooks, are still in common use in modern industrial societies.

There were also a lot of non-material innovations in place in each pre-agricultural society. These were found embedded in the habitual patterns of labor (such as hunting methods), in community defense, in the manner of settling disputes, and so on. So the societies that first entered upon agricultural lives were not blank slates; they brought a lot of embedded innovations with them. Still, the rate of accumulation of innovations shifted rapidly upward when agriculture was invented. This shift is called the Neolithic Revolution.

At each of the seven origins of civilization, agriculture was present in the region long before the first city appeared there. In some cases, such as China, agricultural villages existed for hundreds of generations before the first known city was built. The appearance of agriculture is therefore a key issue in understanding how we ended up surrounded by our artificial world of innovations. It's also a frustrating one, because there's no consensus on the question of how agriculture first appeared. The seven primary civilizations all occurred within regions where agriculture had spread from just three points of origin: a wheat-based system from the Levant, a millet-based system in China, and a corn-based system from Central America. Agriculture also originated in other places without ever leading to any civilizations, for example in sub-Saharan Africa and New Guinea.

The very first appearance of agriculture occurred 600 generations ago in the Levant region, not far from Mesopotamia, where the first civilization would eventually arise. Keep in mind, though, that most of the world's arable lands didn't feel the plow until much later. For a very long time, both agriculture and civilization were merely local phenomena, found in just a few small patches of the world. In fact, large swaths of the planet continued to be populated by band cultures that had neither agriculture nor cities, right up until the age of European exploration got started, just 25 generations ago.

One huge effect of the innovation of agriculture was to concentrate the most precious resource of the time – food – into storage facilities such as granaries. It also concentrated food-production resources, especially upon patches of arable land. A community that gets its food from hunting game needs access to a vast expanse of forest or grasslands, but one that grows grain needs only small, precious plots of good soil, where labor can be invested intensively. I point out this aspect of agriculture because we're about to see that this tendency to concentrate resources is the most obvious thing that agriculture, tax and cities all have in common.

Perhaps we can't exactly prove that civilization appeared as a result of the progressive concentration of resources, but those two things certainly happened at the same time. Here's one possible clue in that regard: None of the seven original civilizations appeared in a culture that had animal husbandry but no agriculture. Animal herders, or pastoralists, need much more land for their livelihood than grain farmers, so they don't naturally tend to concentrate resources into one place as much as farmers do. Pastoralists have founded many civilized societies, but never an original civilization. To give a few examples, the Incas, Mongols, and Arabs all began as pastoralists and went on to found vast empires. In every case in which a herding culture (whether village-based or nomadic) has made the transition to city life, it has done so through conquest of, or fusion with, agriculturalists.

How does the innovation of agriculture prepare a society for civilization? In other words, how does it predispose a non-civilized society to perhaps one day start building cities? The problem with this question is not that we don't have an answer... the problem is that we have dozens of answers, and a long history of vehement squabbling over which one is correct. Fortunately, we can proceed with the project at hand in this book without having to know which answer is right. We only have to observe that the following steps can occur in the real world, and have in fact occurred several times:

1) People in a certain region begin growing crops.

2) The region experiences a concentration of valuable commodities (such as food) and the means of producing them.

3) The rate of innovation goes way up. We call this a Neolithic Revolution.

4) The people in the region begin building cities.

I leave it to the reader to fill in the spaces between those four statements, using the works of Spencer, Marx, Weber, Childe, Steward, Service, E. O. Wilson, or anyone else. There are some reading recommendations in the Notes at the end of this book.

The custom of taxation is one of the innovations that occurs after the appearance of agriculture and before the appearance of an original civilization's first city. The question of what, exactly, tax represents is perhaps the most crucial element in the debates over the development of civilizations. Some authors (for example, Karl Marx) regard tax as value extracted at swordpoint from laborers to enrich a parasitic aristocracy. Others (for example, Elman Service) believe that tax is part of a mutualistic, "evolutionary" process in which complex administrative services develop, to the benefit of more or less everybody, allowing the creation of roads, new industries, etc. Once again, I have the good fortune to be able to avoid saying anything about that debate. For our purposes, the only thing that matters is that, one way or another, agriculture and tax are key innovations in the transition of clan-based hunter-gatherer societies into civilizations. There are probably a number of other such innovations as well, such as certain modes of military organization. But agriculture and tax share in common their tendency to concentrate value in particular places. Those places are where the first cities appear.

Next question: What do we mean by 'city'? The first cities didn't look much like modern Shanghai or New York. In what sense were they different from villages? Since their appearance was, by our chosen definition, the beginning of civilization, this question needs a clear answer. Dictionaries are going to let us down again, telling us that cities are "large, permanent human settlements," and similar phrases. What do we mean by large? Lots of American 'small towns' are bigger than Athens in its Golden Age. And we all know that nothing in this world is permanent. Carthage was perhaps the greatest city of the Mediterranean until the Romans got tired of having wars with it, and erased it right off the map. So much for permanence.

Let's see if we can come up with a better description of a city. Although the first cities didn't look much like today's cities, there are some reasonably sharp distinctions between them and the villages that preceded them. To distinguish a village from a city, archaeologists look for certain characteristic features, two of which are of particular importance. One is a population in excess of some minimum size, which can be anywhere from 1000 to 20,000 residents, depending on which expert you ask. The other characteristic feature is a surrounding defensive wall. These walls are so important in defining the first cities that a case can be made that they are the third key innovation leading to civilization, along with agriculture and tax.

When an archaeological dig reveals a settlement that could be called either a village or a city, archaeologists are prone to shrug and label it a "proto-city." This term is most often applied to a large and apparently prosperous village that was found in a time and place that later produced true, unambiguous cities, but which itself had both rural and urban features. As a general rule, if it had a wall around it, it was a real city. The tough cases tend to be big, rich, long-lived settlements with no protective wall.

There's a good reason that the wall is so important. Agriculture concentrates precious resources, as we've seen, and tax further concentrates them into even bigger hoards. If a group of people is living largely off the concentrated tax resources they receive (or extract) from nearby farmers, their homes tend to fill up with precious assets of various sorts. What could be a more tempting target for bandits and raiders than a dense cluster of rich homes?

At some point during the historical process of wealth concentration, it becomes cost effective to build a wall around the houses, and post armed guards to keep a lookout. One of the main expenses of a city wall is the manpower needed to keep those guards posted. An unmanned wall is next to useless, as it can be easily climbed or breached. So, between all the materials and labor needed to build the wall itself, and the financial commitment to maintain armed guards in perpetuity, a community has to be pretty well off to afford a wall.

Consider the original historical case, as an example. After the first innovation of agriculture, 600 generations ago in Mesopotamia, there was a steady accumulation of innovations among local villages in terms of food cultivation, labor management, armaments, and many other aspects of day-to-day life. By 400 generations ago, there were distinct farming villages in the area. There were probably earlier villages as well, but no trace of them has been found yet. But it wasn't until 270 generations ago that the accumulation of innovations and resources in the region led to the world's first walled city: Uruk. That's 330 generations – over seven thousand years – that passed between the first precious granaries and the first true city. During all that time, the idea must have occurred to people over and over to put a manned protective wall around their village. They just couldn't afford it.

A good example of an earlier, unwalled economic center was Çatal-höyük. This was a thriving proto-city about 400 generations ago in Anatolia, part of the same original agricultural region as Mesopotamia and the Levant. It consisted almost entirely of a dense cluster of well appointed homes... a bandit's dream. The idea of protecting this wealthy cluster of buildings from raiders with a manned wall must have been obvious, but no wall was ever built. Presumably, it was just too expensive. If they had managed to get a tiny bit richer, it seems likely that they would have built a manned wall to protect their wealth, and Turkey (rather than Iraq) would have been the cradle of civilization.

At any rate, once a community's hoard of precious assets (such as grain taxed from farmers), grows large enough, the people do eventually build a manned wall to protect their homes. The immediate result of doing so is that the walled settlement's value rises dramatically. The protected space inside the wall suddenly becomes intensely desirable as a home for the elite, and as a marketplace site, and as a place to keep armories, granaries, treasuries, and so forth. The erecting of a fortifying wall is likely to give a dramatic boost to any socioeconomic center that manages to get rich enough to build it in the first place. The wall instantly raises property values by a huge margin, converting a cluster of villages into a thriving, bustling city, practically overnight.

As soon as the first cities appeared in any given region, there was another great increase in the rate of innovations, even more intense than the one that followed the discovery of agriculture. This transition is known among archeologists as the Urban Revolution. Notice that, once again, the leap forward in terms of rate of innovation happened at the same time as a big increase in asset concentration. That was definitely true after the innovation of agriculture, and probably true after the innovation of tax, and again *definitely* true after the innovation of city walls.

So, that's where civilization came from. The next really conspicuous increase in the rate of innovations wouldn't come for about 260 generations, when the Industrial Revolution kicked off, just twelve generations ago. In the meantime, innovations piled up fairly steadily in each of the separate civilization centers around the world.

Perhaps the most crucial innovation during that whole long period was imperialism, which took a surprisingly long time to develop. The idea of imperialism is for one city to gather up other cities and their lands as assets, either by taking possession of them outright (like one person enslaving another) or charging them tribute (like one person taxing another). The first multi-city empire, Akkad, didn't appear until 200 generations ago, in 2350 BC, fully 1600 years after the founding of the first city.

Innovative improvements in administration gradually converted multi-city empires into kingdoms. These administrative innovations eventually came to include elected parliaments, public school systems, international espionage agencies, etc., as the kingdoms transformed into the republics that currently house most of our global population. The exact nature of the social processes that caused all those administrative innovations is, once again, a matter of unending debate. All that matters for our purposes is that, one way or another, all sorts of innovations piled up during those centuries, and they're still with us today. We're practically swimming in them.

3

THE ROAD TO HAPPINESS

We've seen that innovations are embedded in the materials and labor practices all around us. We've also seen that innovations have been gradually accumulating for a very long time, and are an essential difference between our modern, civilized lives and the lives of Paleolithic hunter-gatherers. From one point of view, we can even say that they are *the* essential difference.

Now it's time to take a closer look at one aspect of innovations that we've only glanced at so far. Embedding innovations in assets, materials and labor practices adds value to those things – value that we can put a price on. Because of that, the long-term accumulation of innovations, which has been going on since before the discovery of agriculture, is an essential source of economic growth for societies.

The fact that innovations can pile up in a society over time, increasing the value of pre-existing items, makes long-term economics something very different from a "zero-sum game." The stakes get higher and higher over time, because even if the same land, labor capacity and materials are passed from generation to generation, they keep increasing in value as innovations are added. This view of social progress was advocated by the Harvard economist Joseph Schumpeter in the 1940s.

At the other extreme, Karl Marx's views of socioeconomic development are usually regarded as strongly opposed to the view that technological progress and innovation are engines of real economic growth. But even he once wrote: "The windmill gives you society with the feudal lord; the steam-mill society with the industrial capitalist." Although he probably meant something like, "the new boss is the same as the old boss," we all know that a feudal society and an industrial society are *really* different places to live. The idea that technological innovations are the basis of the difference between them is a remarkable thing for Marx, of all people, to have observed.

Saying that a society's wealth is largely defined by the size and quality of its hoard of innovations is nearly identical to saying the following: A nation doesn't get a strong economy by simply sitting on lots of natural resources and having a big labor pool. There are plenty of miserably poor nations that have both of those things. The resources and the labor pool must undergo development in order to have high value. That process of economic development is precisely the process of embedding lots of high-quality innovations into the nation's assets, workers and culture.

This use of the term 'developed' to describe a wealthy, innovation-packed society has a big advantage over the older term, 'advanced'. When we say a nation is 'developed' in this sense, we needn't be speaking teleologically, because we're not claiming the society has advanced toward some goal. We're don't even have to claim that the development was all for the good, nor that all of the development was rational. When we say a society is highly developed, our choice of words only commits us to saying that the society is a long way from its original condition, which was the clan-based hunter-gatherer band. The distance between that ancestral condition and any modern society can be measured by the quantity and quality of innovations that it has embedded.

Of course, we can also say that a society's wealth consists of its accumulation of valuable resources, including not only its land and material resources but also its labor pool. But the value of these things depends intimately on how many innovations have been embedded in them. We've already seen the many ways in which labor skills and material assets can gain value due to the accumulation of innovations. Land, too, rises in value as innovations build up. The biggest factor in determining the price of land is its location, and the most critical aspect of location in terms of real estate value is how close the land is to an innovation-rich hoard, such as a city center.

The personal experience of life for an average individual living in a highly developed society is very different from that of an individual living in a poorly developed one. Again, as with the society as a whole, this effect of development can be expressed either in terms of wealth or in terms of embedded innovations, and the two views turn out to be nearly identical. The advantage of speaking in terms of wealth is that it's easier to measure. The advantage of speaking in terms of embedded innovations, on the other hand, is that it shows the link between people's motivating drives (which cause them to innovate one thing rather than another) and the development of nations. The fact that the two views are so tightly linked will make our job here a lot easier. It's not easy to count a society's embedded innovations, but simple economic indicators, such as GDP, can give us a good approximation of how many have accumulated.

GDP, gross domestic product, is the total amount of money-valued productivity in a given nation during a given year. Most of this productivity is in the form of labor that adds value to materials, but productivity can also be in other forms, such as money-valued services (for example, cleaning someone else's house... but not your own).

For the purposes of this book, per capita GDP is a much more important number than GDP itself. Per capita GDP is the amount of the nation's money-valued productivity that is accounted for by the average citizen – in other words, it's the GDP divided by the population size. Since that productivity is set at a price, the labor is pretty much always paid for. The pay comes in the form of paychecks, retirement funds, unemployment insurance, perquisites, and so forth. Exceptions exist, such as slavery and the traditional role of the housewife, but they're increasingly rare in the modern world. Per capita GDP is quite a good indicator of the average citizen's day-to-day experience of prosperity or poverty, in most societies.

The difference between a nation's GDP and the same nation's per capita GDP can be important, and we're really only going to be concerned with the latter. The GDP of a nation is mainly useful in understanding how the country as a whole behaves, not for understanding what life is like for the actual people who live there. For example, Malaysia has a slightly bigger GDP than its neighbor, Singapore, and that matters a lot when diplomats from those two countries sit down to negotiate over pretty much anything. But if you walk through the streets of Malaysia and those of Singapore and have a look around, what you'll mainly notice is the extreme difference in socioeconomic development between them. Singapore looks and feels far more prosperous than Malaysia even though its GDP is lower, because its per capita GDP is three times higher. For our current purposes, we're concerned with human experience rather than the competing careers of nations, so per capita GDP is the measure we need.

The accumulation of embedded innovations can have profound impact on people's lives, and the source of that impact isn't always direct and obvious. In fact, the indirect and general impacts can be more important than the direct, immediate ones, in many cases. For example, the ambient, background density of embedded innovations in a society is often the main thing that determines a person's economic quality of life. That's surprising, because we might imagine that each person's set of skills, together with his or her social class, would determine his or her economic status. But in fact, these things are often fairly minor factors, while the nation's level of development as a whole is the main thing determining each person's level of prosperity. A specific example will make this more clear.

Imagine that we're watching an American construction worker on a site where a crew is building a residential home. The foundation is poured in concrete, and then a wooden platform called a floor box is created on top of it. The wall frames are then built one by one, along with the truss frames that will hold up the roof. These pieces of wooden frame are generally constructed somewhere away from the actual building foundation, nailed together lying flat on the ground. In fact, it's fairly common for the wall and truss frame units to be built at a factory and trucked in, pre-fabricated. Once these modular pieces are ready, they are lifted into place on the floor box, then joined with steel fasteners that range from nails fired out of nail guns to large galvanized steel plates and ties.

And now, imagine we're watching a construction worker on a residential building site in Indonesia, erecting a building that's about the same size as a typical US home. In Indonesia, that might be a small apartment building. Raw wooden beams show up on trucks and the worker and his crewmates spend long hours sawing and planing them by hand, without power tools. To minimize the use of expensive steel fasteners, they work with hand chisels to fashion mortise-and-tenon joinery for the larger frame pieces, which can then be assembled "stick by stick" (as the expression goes). The whole process of framing the building may extend over an entire season and incorporate vast numbers of man-hours.

At the end of each week, the American worker and the Indonesian worker collect their pay. The American worker receives about eight times as much per hour as the Indonesian worker, in real terms. Let's try to figure out why. One thing it surely cannot be is a matter of class difference, because both of these people are members of the same working class, although in different nations.

In Chapter One, I pointed out that the labor value of manual work done by surgeons in the US is valued at sixteen times as much as that of 'unskilled' workers. We saw that most of that difference can be explained by the long training process that embedded the surgeon's labor with a huge body of medical innovations. But that observation isn't going to help us much in trying to understand the difference in wage between the construction workers of Indonesia and the US. An Indonesian construction worker may have spent as many hours in training and apprenticeship to learn the necessary skills for his craft as the American did – he has a *different* set of skills, but not an easier one.

Another approach we might take in trying to explain the huge difference in pay is to say that the difference is a matter of historical fortunes – the good fortunes of American history and the bad fortunes of Indonesian history. There is surely some truth to that, but it's still not nearly the whole story.

A big part of the difference in wages is in the quantity and quality of innovations embedded in the two workplaces, rather than the innovations embedded in the workers themselves. The power tools, the steel joiners, the economy of scale due to factory-made lumber and even factory-assembled frame components, plus the more efficient assembly plan and workflow that come with these things, all make it possible for a single worker with the same amount of training to generate *much* more productive output per day, in a more developed nation. A house frame that goes up in a few days in the US might literally take months to assemble in Indonesia. It's not that the laborers of one country are intrinsically better workers than those of another, nor even that more effort has been invested in their training; the issue is that one society is more thickly saturated with embedded innovations than the other. As a result, an hour's work simply gets more accomplished.

Here's what that shows us: Even though per capita GDP is only a rough measure of a nation's level of development, and even though it lumps everyone in the nation together, it's still quite a good indicator of the day-to-day experience of prosperity of the average citizen in most societies. Nations with higher density and quality of embedded innovations have higher per capita GDPs, because all of their exchanges of money, goods and labor are happening in a richer economic environment. In a developed nation, where the per capita GDP is high, everyone is swimming in a thick soup of embedded innovations, and this empowers their every action. Workers generate more productivity in a day (and get paid accordingly), consumers get more value out of their purchases (and spend more to get it), and everyone's recreation time is full of extravagant choices.

Since per capita GDP isn't too hard to measure, it makes an excellent indicator of how richly embedded with innovations is any given society.

We're ready, then, to answer the question: How quickly are innovations piling up around us? Now that we've established that per capita GDP reflects this density of embedded innovations, we only have to determine how much richer we are today than we were in previous generations, and we'll have a robust estimation.

Since the dawn of the Industrial Revolution in 1760, the average person on earth has become ten times richer, in real terms. By "real terms," I mean after adjustments for inflation, for the different values of money in different countries, and so forth. Worldwide, we humans have gotten ten times richer in just twelve generations. The actual historical estimates are shown graphically (with sources) near the beginning of Appendix One, in Figure A1.

The typical citizen of today's developed nations may not *feel* rich, but if he or she met face-to-face with the ghost of a citizen of the same nation and social class from 1760, the ghost would be in awe at how much we modern people take for granted. We live to be eighty years old! Our kids don't routinely die of cholera or typhoid! We've even got light bulbs, not to mention microwave ovens and the Internet. Admittedly, the wealth in modern societies is not evenly distributed... but it was even less so in 1760. Since then, among other things, colonial rebellions have reduced the global empires of England and Spain to rump nations, and revolutions have brought down three of the biggest aristocracies: those of France, Russia and China.

The fact that per capita GDP has risen steeply, worldwide, since the Industrial Revolution simply reflects the fact that embedded innovations have accumulated quickly since then. But what about before 1760? As the previous chapter explained, the process of accumulation of embedded innovations didn't start twelve generations ago; it's been going on for hundreds of generations. The process has accelerated in recent generations, true, but it didn't begin with the invention of mass-production factories and steam powered engines.

Direct measurements of various nations' GDPs before 1760 are not easy to do, and the results of historical economic studies that try to estimate these old, historical GDPs are contentious. But we can get an indirect indication by using a little deductive reasoning. We may not have records of every financial transaction from (say) the year 1400, but we do have fairly reliable estimates of world *population* back then. That turns out to be useful.

The human population was once strongly linked to prosperity, because prosperity was mainly a measure of the resources needed for survival – such as food. For typical people back then, the most likely causes of death were closely linked to the effects of poverty. These included not only malnutrition but also disease borne by unsanitary water sources, infectious diseases such as smallpox and malaria, bacterial infections, and so forth. The rich weren't immune to these blights, but they were far less vulnerable than the poor.

As prosperity begins to rise in a less developed nation, death rates go down sharply among adults and especially among infants, so life expectancy rises, and the standing population of the nation increases. If the per capita economy remains strong, then this new and higher population density can be sustained. Thus, a historical rise in population growth is generally a sign of economic growth. There are two exceptions. One is if the growth is brief, lasting only a few generations before leading to a severe population crash. If that happens, it means the population outgrew its economic base, using non-sustainable resources, as happened in Ireland during the early 19th century.

The other exception is when the nation's per capita wealth exceeds about $5000 per year, as measured in the equivalent of 2015 US dollars. That, of course, describes the condition of most modern nations. We'll look at that case in Chapter Six, and discover that as societies become prosperous beyond that point, they begin to experience declining birth rates, breaking the simple historical relationship of population growth to economic growth. But we're also going to see that no nation on earth was rich enough to experience that effect until around 1700, just fourteen generations ago. Furthermore, the world as a whole didn't grow that rich until the 1940s.

But let's not jump ahead. Given that there was a strong correlation between rising prosperity and growing population in almost every nation in the world, during all of history prior to the Industrial Revolution, what can we say about per capita GDP in those days? We can say that the world population rose for many centuries before 1760 (and in fact is still rising). That strongly suggests that per capita GDP was also rising, during all those centuries. Furthermore, that's hardly a surprising conclusion, because the steady accumulation of innovations during those centuries kept pushing up the value of materials and labor throughout the whole period.

Here's a general summary of what we know about the long-term historical rise of the human population (explored in more detail in Appendix Two, Topic A, and shown graphically in Figure A5). When agriculture appeared, 600 generations ago, the global human population was around five or ten million individuals, and ever since then, human populations have been growing at a roughly exponential rate. There have been setbacks, such as the Black Plague years, but the overall pattern has been one of steady, powerful growth. This long-term growth of global population would have been impossible to sustain during all those centuries if not for a slow but steady climb in global per capita GDP – or what amounts to the same thing: the steady accumulation of innovations. So not only have human societies been piling up innovations quickly since 1760, we've been doing it slowly since about 7000 BC. When you think about how many embedded innovations are around you right now, you may not be surprised to hear that it's taken several millennia to heap up so many of them.

But before we all start patting each other on the back, congratulating ourselves on how clever our species has proved to be, we'd better take a closer look at the question of *why* we've been pursuing this long path of innovation. That will be the subject of the next chapter. The difference between swimming in all these brilliant innovations and drowning in them turns out to be uncomfortably small.

4

THE PLEASURE PRINCIPLE

So far, we've seen that we're immersed in innovations, that they have accumulated over time, and that each of them increases the value of whatever it's embedded in. Now it's time to look at a question that's going to lead us to see all of those things in a radically new light. That question is: What sort of *drive* causes us humans to keep creating all of these innovations?

Let's start with the most obvious observation. People innovate because they're not satisfied, and are making an effort to decrease their dissatisfaction. This observation might seem so bone-headedly obvious that it goes without saying. The reason I want to start there is that, when we phrase matters in those words, we're reminding ourselves that each element of our synthetic human world – each embedded innovation – was born as a result of a motivating drive inside some person's head. No matter how weird, clever or high-tech that innovation might be, the drive that led to its appearance was as natural and biological as the drives toward hunger, sleep or sex.

Innovations are introduced because someone is dissatisfied, and he or she hopes to decrease the feeling of dissatisfaction. That statement is so broad that it certainly includes one familiar motivation: the hope of making money off an invention. A common adage tells us, "Necessity is the mother of invention." But we can't literally mean "necessity" when we say that, because we know that, for any given invention, people were around for many generations before the thing first appeared, and they got by without it. What we really mean is that the desire for something to come along and improve matters was already there, palpable, before anyone figured out what to do about it. If we wanted to be literal (and had a tin ear), we might re-phrase the adage: "Dissatisfaction is the mother of invention."

Although the term 'dissatisfaction' is a bit clunky, it's useful because it covers such a wide range of cases. The word can mean everything from an indisputable human need (such as food when you're dying of starvation) to the most frivolous desire (such as preferring desserts with extra whipped cream on top). The drive to decrease our level of dissatisfaction is al-pervasive among human motivations, characterizing every case that we might call "fulfilling a need" or "decreasing a state of want," as well as those we would be more likely to call "the pursuit of happiness," or just plain "pleasure-seeking." That doesn't mean that all of those things are the same – they aren't. But each of them has, at its foundation, some degree of motivational force that's based on dissatisfaction with the way things are, causing us to do something, or even to go to the trouble of thinking up and executing a plan in order to change things in the future. Such plans might be as trivial as getting up off the couch and going to the refrigerator, or as elaborate as a scheme to conquer Europe, or cure cancer.

Still, it might seem that we could smooth out our discussion by talking about increasing satisfaction rather than decreasing dissatisfaction. Double negatives are ugly, and usually best avoided. But there's a problem with making that switch: We humans have a remarkable capacity to work effectively day and night in our quest to decrease our dissatisfaction, without *ever* becoming satisfied. It's ironic, but true. We can get everything we demand, and yet we may still not achieve any lasting increase in satisfaction. As a result, decreasing dissatisfaction isn't necessarily the same as increasing satisfaction.

The same principle is seen in the development of civilizations over time. Although the means of satisfaction become more and more efficient as the generations go by, there's no obvious tendency for people to become more satisfied. Rather, all those new, more efficient means of satisfaction soon become a 'necessity' – not just something that people want, but something that they believe they need. So many innovations have been added to our list of necessities over the centuries that today we look back at the original, ancestral condition of our species as something utterly alien. We citizens of modern, wealthy societies can't imagine what it would be like to live with no source of food looming anywhere in our future except whatever animals we can kill with stone-tipped weapons, although that condition was universal for our species until a few hundred generations ago. Not only that, we can't even imagine what it would be like to expect that fewer than half of our children are going to grow to adulthood – a condition that was universal among everyone but the richest people on earth just a few generations back. Let's be real... there are plenty of us who can't even imagine getting through the next week without television and fast food.

The human quest to decrease our sense of dissatisfaction thus appears to have a direction and a source of perpetual energy, but no goal or finishing line. In this sense, it's more like a treadmill than a race. But how is it possible for dissatisfaction to have this oddly immortal quality – the ability to renew itself, no matter how successful we are in our fight to kill it?

The answer is clear when we consider the best-studied example of the phenomenon, which is drug tolerance. As a person becomes habituated to an addictive drug, he or she finds that it provides a profound satisfaction, like scratching an itch that the person hadn't even previously noticed was there. But after enjoying this pleasant experience regularly for a few weeks, the user finds that the drug's effects seem weaker, so that ever-higher doses are needed just to obtain the same effects as before. In fact, if the user quits the drug, he or she doesn't go back to the original state of day-to-day life. Instead, there's a feeling of suffering, even though nothing is actually wrong. This feeling comes in the form of withdrawal symptoms, which may include sickness, pain, despair, unbearable craving, or all four at once. The delights offered by the pleasurable drug have somehow made a subtle transition from being a source of satisfaction to being a necessity.

The phenomenon of drug tolerance gives us a clear metaphor for the treadmill we run as our civilized societies embody our endless pursuit of satisfaction. In fact, it's more than a metaphor. Addictive drugs are one of the real conditions of our societies, and are actually just one set of specific examples of a far more general phenomenon. Sure, drug addiction is based on the chemical processes and cellular interactions within the brain... but that's true of *all* our motivational drives. It's hardly surprising that we develop similar 'tolerance' toward every new convenience and pleasure that our societies devise, learning to take them utterly for granted within a generation or two, and feeling we couldn't possibly live without them.

So we're actually off to a good start here, simply in observing that every act of human innovation is an effort by some individual to decrease dissatisfaction. We can even go a little further and say this: *Anything* that causes a person to perform a voluntary act must involve some form of dissatisfaction. When people are completely satisfied they have no reason to act. On the other hand, complete satisfaction is a mighty rare state, except maybe when we're in one of the dreamless phases of sleep.

There's an inevitable logic to the rule that people only act voluntarily in response to feelings of dissatisfaction, but people are complex creatures and our motives are often hard to see. As a result, it doesn't always *look* like every voluntary action is a response to personal dissatisfaction. People are prone to take on heroic burdens, or to go around actively seeking danger and trouble, or even to get perverse pleasure out of pain. No one forces people to act in these odd ways; these are elements of human nature, and we do such things voluntarily. Can these voluntary actions really be explained as efforts to decrease personal dissatisfaction? As it turns out, they can. Let's see how.

People are often willing to accept some voluntary hardships as part of a plan that they expect will yield a future reward in terms of decreased dissatisfaction. When a person intentionally acts to temporarily increase his or her dissatisfaction in this way, he or she is making an intentional investment in short-term hardship as part of a plan to decrease dissatisfaction eventually. We can refer to this sort of investment as **strategic suffering**, and we can define the term like this: Strategic suffering is the voluntary increase of dissatisfaction as an investment in a plan (conscious or unconscious) that's intended to decrease dissatisfaction in the long run.

Here's a simple example. Imagine a fellow pulling some ice cream out of his freezer and opening it up. Then he pauses as he remembers that he's diabetic, has bad teeth, and is getting fat. A clear image of the worst-case scenario of his future self appears before his eyes. He shudders and puts the ice cream away. Somewhere inside him, an inner child is weeping at the cruelty of this deprivation, but he ignores all that and leaves the kitchen to go look for diverting recreation elsewhere.

This person didn't make any conscious calculations as he ran through his set of futile actions with the uneaten ice cream, but the calculations were getting made, nonetheless. He calculated that his overall, lifetime dissatisfaction would probably be *higher* if he ate the ice cream, even though eating it would have sharply decreased his dissatisfaction over the next ten minutes. Somewhere in his brain, he figured that the long-term cost exceeded the short-term benefit, in terms of minimizing dissatisfaction. We don't make choices this way every single time, but our capacity to do so when we want to is a lot of what makes us adult, and human.

So all motivational drives have this much in common: they are always intended to decrease dissatisfaction in the person who's feeling that drive. We might be tempted now to try and replace the awkward term 'decreased dissatisfaction' with a similar and more familiar term from psychology: the pleasure principle. That term was coined by Freud in 1911, and was intended to convey the idea that *all* human motivations have pleasure-seeking (and its twin, which is pain avoidance) hidden at their core. However, he explicitly contrasted the pleasure principle against what he called the 'reality principle', which includes the ability to seek pleasure and avoid pain through indirect routes, deferring gratification and following strategies that require temporary austerities and hardships. The reality principle is what you get when you combine the raw drives of the pleasure principle with the capacity for strategic suffering. The pleasure principle is basically infantile, consisting of little more than a tendency to reach out and grab attractive things, and to recoil away from things that hurt. The reality principle, by contrast, is what adults (usually) do. Our efforts to improve matters aren't limited to just simple attractions and reactions. We can plan ahead, and put ourselves through a certain amount of disciplined hardship to get where we want to go.

Freud's notion of a reality principle allows for much more elaborate behavioral tendencies than a baby shows when it's following the simple pleasure principle. And yet, the reality principle doesn't require the addition of any new drives or sources of motivation beyond those of the pleasure principle. In fact, it's nothing more than an extension of the pleasure principle to include indirect, strategic approaches to the satisfaction of those drives. It just adds reason and discipline to the kit of faculties a person has on hand to obey the drives created by the pleasure principle.

As far as terminology is concerned, we can see now that the term 'pleasure principle' is not enough to explain the complexities of human motivation and behavior. On the other hand, the term 'reality principle' covers every case in which a person acts to decrease their feelings of dissatisfaction, no matter how indirectly or subtly they do so. The only reason I prefer to speak in terms of 'decreased dissatisfaction' is because the words 'reality principle' are so vague that they give almost no hint of what Freud was talking about. The words 'decreased dissatisfaction', on the other hand, are pretty much self-explanatory.

In addition to cases of strategic suffering, there's another common case in which people seem to defy the plain logic of acting to decrease their dissatisfaction. We often see people intentionally throw themselves into risky or downright painful situations, with no hope of future reward. A lot of 'extreme' sports such as mountain climbing have this quality, as do many high-risk jobs such as that of a commando or a fireman. The majority of people who select these activities will say that they prefer such a life, and that they enjoy the challenge, the thrill, etc. If we can take them at their word, then our difficulty in understanding these choices vanishes. These high-spirited activities are another form of pleasure, even if they involve painful experiences and risk. A peculiar twist on such pleasures is seen in the fairly widespread phenomenon of sexual masochism, in which pain becomes a clear and straightforward source of pleasure for some people.

There's a third case in which we might suspect an exception to the principle that decreasing dissatisfaction is the universal drive behind our voluntary actions. Some religious zealots have been known to take on lives of nearly unmitigated suffering in order to achieve spiritual goals. It would sound odd to propose that such a person is engaging in 'strategic suffering', because he or she has no plan to minimize overall lifetime dissatisfaction. Just the opposite: some ascetics seem bent on getting as much dis-satisfaction as life can provide. The great majority of people who live such lives will say that they expect to benefit in the long run nonetheless – but they believe that 'the long run' extends beyond the time of their physical death. We don't need to make any opinion, one way or another, as to whether they're correct about that. All that matters for our purposes is that they believe it, which means that they're performing the same kind of calculation that made the man in the story, above, put away his ice cream.

So there's no reason to imagine that any of these apparent exceptions are real violations of Freud's reality principle, after all. In the big picture, we humans always act in ways intended to make ourselves less dissatisfied. The *only* time our voluntary actions increase our dissatisfaction is when they are elements of a plan to decrease it eventually.

Now we should dig a little deeper and try to find the source of our perpetual dissatisfaction, given that it's the drive behind all of our actions. Since we're confining ourselves to natural explanations, we quickly see that the immediate source of the drive has to be some sort of neurological circuitry found in the brain. By 'neurological circuitry', I mean the pathways of information in the brain that consist of brain cells and the chemicals that they use to communicate with each other.

The ways in which our experiences are linked to the actions of these neurological circuits are mostly unknown, but a few of the links have been worked out quite well. Over the past sixty years, science has discovered a lot of the essential circuits that increase and decrease dissatisfaction. These are generally referred to as 'reward' and 'punishment' circuits, since they tend to drive our behaviors through the pleasure principle. We aren't abject slaves to these givers of reward and punishment, but they are always there to nag us, and we always take them into account on some level when making our decisions.

Only the simplest examples of these neurologically based drives have been fully described by science. Still, we can say this much: *Some* sort of neurological reward/punishment circuitry *must* exist even for the most sublime and abstract forms of pleasure, pain and dissatisfaction. We don't perform voluntary actions if we're not dissatisfied, and the source of that dissatisfaction has to be somewhere in our brains. It's true that many things outside the brain contribute to our feelings, including our senses, hormones, interactions with others, and so forth. But it's up to the brain to organize and interpret all those inputs in order to construct our perceptions, emotions, thoughts and consciousness. The only behavioral drives that originate outside the brain in humans are simple reflexes (and these, of course, are not voluntary actions).

Simple, basic pleasures such as hunger, thirst, sleepiness or the sex drive are easy to explain, right down to the nuts-and-bolts level of the brain's neurological circuitry. Let's take hunger as an example. The exact details of this story won't be crucial to us here, but the case should nicely illustrate the principle.

Hunger works like this. A large nerve called the vagus connects the brain directly to the stomach, along with most other organs in the torso. When the stomach is empty, the vagus sends nagging signals straight to the brain. This stimulates the brain's hunger center, which is found near the middle of your head and straight back from the center of your nose, in the paired lateral regions of the hypothalamus (Figure 3). The brain cells of the hunger center respond to these various cues by producing a signal molecule called orexin, which is effectively pure, chemical hunger. There are only a few thousand cells in the hunger center (out of the total population of 100 billion brain cells), but they send out their signals through a network of long extensions, delivering them to regions scattered all over the brain. Their power to drive our behavior can be, shall we say, considerable.

If all of that is happening (in other words, if you're good and hungry), then when you begin to eat, the hunger center will reward you for your obedience by sending a stimulating signal to your pleasure center. The

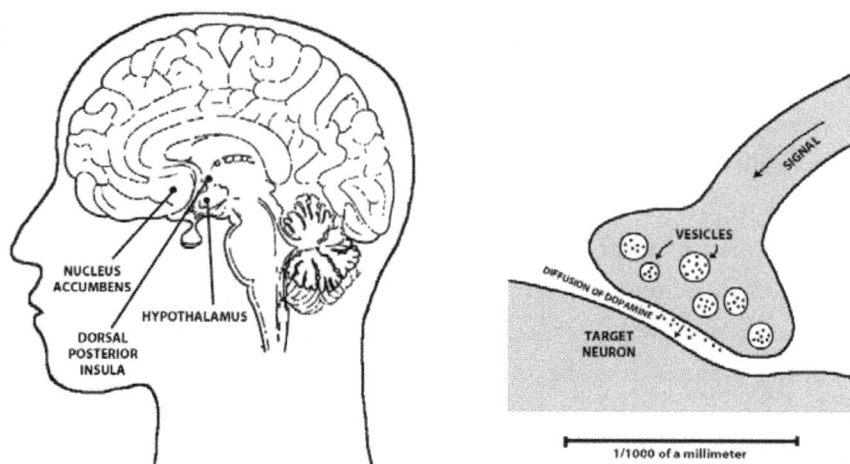

Figure 3. Chemical pleasure.

The picture on the left shows the location of some brain regions described in the text, including the pleasure center: the nucleus accumbens. The picture on the right shows an axonal knob, the part of a brain cell that releases signal molecules, such as the dopamine shown here. These dopamine-containing knobs are found on the ends of cell extensions that arrive at the nucleus accumbens from many distant parts of the brain. A signal arrives at the knob, where the dopamine is kept stored in hollow balls of membrane, called vesicles. The vesicles respond to the signal by dumping dopamine onto the surface of one of the cells of the nucleus accumbens. Our experience of this delivery of dopamine is what we call 'pleasure'.

hunger center does not, in itself, have the power to make you feel good – it can only make you desire or not desire to eat. But each of the separate brain centers that controls a basic motivational drive (hunger, sex, thirst, sleepiness, etc.) has the power to cue the pleasure centers, as a reward for carrying out the motivational center's demands.

Pleasure is controlled by interactions among several brain regions, but the key pleasure center is called the nucleus accumbens. It's found conveniently close to the hypothalamus, where most of the motivational drives are located. The cue that tells the nucleus accumbens to dole out rewards (for example, because you were hungry and you got yourself some food) consists of a signal molecule called dopamine, delivered into the pleasure center on long extensions of brain cells that live in other regions of the brain. As shown in Figure 3, the dopamine is usually stored in small chambers called vesicles near the tip of these brain cell extensions. A signal travels the length of the extension, and causes the dopamine to be dumped onto the surface of a target brain cell inside the nucleus accumbens. Our experience of this chemical stimulation is what we call 'pleasure'.

The most remarkable aspect of this story is that although our many motivational systems may be quite complicated, the pleasurable aspect of each one of them turns out to be simple... and the same in virtually every case. That's surprising because, even if we're only talking about one motivational drive, such as hunger, there are many pathways of nerves and hormones and so forth distributed around the body and brain that contribute to the sensation. Furthermore, we're able to experience many varieties of the drive. I might be feeling "hungry for pizza" or "so hungry I could eat a horse." Certainly the sensation of acute hunger that I feel after missing two meals in a row is very different from the chronic hunger following an intense month-long diet, even if both are equally severe. But satisfying any form of hunger will cause a brief release of dopamine in the nucleus accumbens, and thus give us our reward, namely: pleasure.

Separate from the pleasure system in the brain, we also have a group of regions that give us the experience of pain. The primary pain center is the dorsal posterior insula, again found not far from the hypothalamus. It's stimulated by pain-signal molecules (such as substance P and neurokinin A), delivered on long cell extensions, much like the stimulation of the nucleus accumbens with dopamine. The effect of these chemicals is what we experience as 'pain'.

There's also a set of circuits that create a strange interaction between the brain's primary pleasure and pain systems. These use a set of signal molecules called endorphins, such as beta endorphin and enkephalin, which act to dampen the punishment action of pain circuits. This can be crucial in emergency situations where intense pain might distract someone from reacting to the circumstances. For example, let's say it's 10,000 BC and you've just been clawed by a leopard. That's a bad moment to just lie down on the ground and howl... you really should be running as fast as you can, and do your howling later. Endorphins defer the pain for a little while to help you have a fighting chance in such situations. In addition to killing pain, the endorphins also create pleasurable, euphoric states of mind – an effect similar to that of opiate drugs. This may or may not represent a second type of primary pleasure system, and there is an ongoing debate over whether endorphins create pleasure directly, or by causing dopamine release.

All the observations in this chapter so far apply to the process of innovating, just as much as they apply to any other voluntary human activity. When a person innovates, it's always part of a plan to decrease his or her dissatisfaction. And yet, when we're explaining the process of innovation from this neurological point of view, we find that Freud's pleasure principle is pretty much useless. When we humans are following the pleasure principle, we operate instinctively, without planning ahead, grabbing at whatever looks good at the time. That process rarely leads to innovation. Innovations come from our ability to follow clever, indirect paths in our quest to satisfy the reward systems in our brains, and to cook up plans for improving access to those rewards. Almost always, such plans involve an initial investment of hardship and labor – a period of strategic suffering. Our uniquely human capacity for planning (and for exercising the discipline to execute our plans) becomes an advantage precisely when there's a barrier between the current state of things and a future state of decreased dissatisfaction. Situations like that require that sacrifices must be endured in order to get to the greener grass on the other side.

Here's what we've managed to establish, then: Every act of planned innovation originates from a drive to decrease dissatisfaction, which is driven by the reward/punishment circuitry of the brain. If that conclusion seems so obvious that it was hardly worth spending so much time on, then so much the better. It's going to lead us to some strange conclusions.

*   *   *

Before wrapping up our investigation of the motives behind innovation, we should give a few moments of special attention to one of the most persistent delusions that we humans are prey to, when we try to understand what motivates us. I'll present this through a short parable, which will illustrate the principle of strategic suffering and then expand it a bit. That expansion may help us to perceive a very common misconception.

This is the story of the Three Brewers. Once upon a time, three people independently came up with the idea of brewing beer, in three parts of the world where alcohol had been previously unknown. (Let's say they were all male, to save a little space on pronouns.) The First Brewer came up with a teleological plan intended to exploit his innovation in order to decrease his dissatisfaction. His plan was that he would set up a home brewery, drop pretty much everything else in his life, and just drink beer. For a while, this worked out perfectly, but in the long run it didn't go so well for him. After a few months of steady drinking, he could no longer afford to repair his house or buy food. Finally someone came in and burglarized his home while he was passed out on the sofa, and took away his beer, his supplies, his equipment, everything. He was left ruined, sober, and sporting a hangover of epic proportions.

The Second Brewer came up with the same idea in another part of the world, but his plan to exploit his invention involved a scheduling system. The schedule obliged him, at regular intervals, to stop drinking beer and apply self-discipline in the form of periods of austere, sober activity in order to deal with his life's day-to-day business – in other words, sometimes he would so some work. That work included securing his house and arming himself against burglars. By adjusting his schedule a little at a time, he eventually found the perfect balance between maintaining his life's infrastructure (such as stocking up on food), maintaining a state of military preparedness (such as checking the locks), and indulgence in the fruits of his labors (in this case, beer). This Second Brewer lived a fine life. He was the living embodiment of Freud's reality principle, employing the tactic of strategic suffering in a prudent fashion that minimized his overall dissatisfaction on a lifetime basis.

The Third Brewer came up with the same beer-making innovation somewhere else in the world, and also came up with a strategic austerity plan similar to the Second Brewer's. But then he decided to extend that plan, pushing his production schedule to the limit and foregoing every opportunity to relax and enjoy. He devoted all of his time to business and military pursuits, so that he could not only brew his beer and protect it from his neighbors, he could sell it off and get rich. Eventually he had more money and weapons than anyone around. After a time, he bought out most of his neighbors so that he could expand the borders of his property. Any neighbors that he couldn't buy out, he robbed at gunpoint. His life never contained a moment's peace, but he died rich and powerful.

So, here's a question: Why did the Third Brewer choose to live that way? We all know that it's a fact of life that people and groups of people often behave in such fashion. Productive and military labor is not all devoted to maintaining the minimum infrastructure and security needed to maximize the time available for exploiting pleasurable pursuits. Often, entire nations will go to war, exposing their citizens to immense hardship and risk, in preference to a perfectly viable alternate plan consisting of staying home and advancing their infrastructure, their security measures, and their luxuries within established borders. So again we ask the question: Why?

The most straightforward answer is that this is a calculated strategy to enrich the aggressors with land and assets, and perhaps also with labor resources and valuable innovations. The fact that the strategy sometimes backfires, ruining the aggressor, doesn't argue strongly against this view. If that interpretation is right, then invasive, warlike cultures are just another approach to the reality principle – strategic suffering endured as part of a long-term plan to decrease dissatisfaction.

The strongest voice arguing against this view was that of Nietzsche, and his influence is still widespread, even among a lot of people who think they're opposed to his opinions. Particularly, many of us have a persistent tendency to think of civilizations as emerging and enduring through some human drive to conquer and dominate. We may hear the story of the Third Brewer and shrug and say, "Well, that may have been a fun and exciting life... at least he was never bored." Almost everyone has some feeling that the human desire to rise to a greater level of power is a primary natural drive, much as hunger is a primary drive. That point of view was systematized by Nietzsche, who called the drive a "will to power."

Does the brain contain a drive toward empowerment, which rewards us with dopamine when we advance socially (for example economically, politically or militarily)? Neither a 'yes' nor a 'no' answer would please Nietzsche. If we say 'no', then the Third Brewer and similar strategic aggressors are really just advancing a plan of strategic suffering that they hope (perhaps unwisely) will one day give them so much luxury that their lifetime sum of dissatisfaction will end up lower, despite all the hardships endured along the way. On the other hand, if we say 'yes', then we're saying that this aggressive tendency is itself a primary pleasure (a joy in bullying, or even pure bloodlust), which the Third Brewer pursues as a form of self-indulgence. We'd be claiming that the Third Brewer preferred to compete with and attack his neighbors merely because it was a form of pleasure that he preferred over the quiet pleasures at home – in other words, just for fun. Nietzsche would certainly never allow either of those two interpretations. As I mentioned in the Introduction, this makes it impossible to interpret Nietzsche's idea of the 'will to power' in any way except as a claim to have discovered some supernatural force.

Incidentally, Nietzsche also wanted to avoid concluding that all our struggles and toil are merely intended to keep our genetic lineage (children, grandchildren, etc.) alive in future generations. That family-oriented view is exactly what Darwin had claimed, a generation earlier. Nietzsche rarely commented directly on Darwin, and many scholars believe he never actually read Darwin's books. Still, he is often considered to have understood Darwin's emphasis on reproductive success – and to have disagreed vehemently. This is hardly surprising, since Darwin's views were confined to reasoning from efficient, natural explanations, much like the line of thought that we're pursuing in this book. Nietzsche, like the religious thinkers that he opposed, was unwilling to accept that sort of limitation.

At any rate, as I mentioned before, many of us tend to think of the history of civilization as a struggle for power and dominance, even if we couldn't care less about Nietzsche. I wonder if this might be the fault of historians. Historians over and over again present the basic 'facts' of history as a long series of battles, coups d'état, political debates among leaders, squabbles within royal families, and so on. That seems odd, since a very large majority of people in both the present and the past have experienced those events only as dim echoes, while their real experience of life has mainly consisted of the immediate conditions in their households and workplaces. History books are full of Third Brewers, and they make ripping good tales. But real life is mainly filled with Second Brewers, because their dull, dependable strategy is the one that truly predominates in the world.

As we come to the end of Part I of this book, let's review what we've established:

1) Satisfaction and dissatisfaction are always neurologically based.

2) Despite our willingness to undergo strategic suffering, the tendency to decrease dissatisfaction is the ultimate basis of all our motivating drives.

3) In order to decrease our dissatisfaction (and only for that reason), we innovate.

4) Innovation is the basis of the rising prosperity of civilized societies, and that has been the case for a very long time.

These four points establish a link between the neurological circuits of our brains and the economic growth and cultural development of civilizations. That link will serve as the basis for Part II, as we have a look at the phenomenon of falling birth rates in wealthy nations. In fact, it will turn out to explain the world's ongoing fertility crisis quite well. That will allow us to make some confident predictions about where our falling birth rate is going to take our species, in the not-so-far future.

Part Two

CULTURE VERSUS EVOLUTION

5

ON HAVING A LOT OF CHILDREN

We've seen that, for the past six hundred generations, human societies have been accumulating innovations. We've found that this process amounts to a long-term increase in our ability to directly or indirectly stimulate the regions of our brains that decrease our dissatisfaction. Those regions include pleasure centers and centers that repress pain, as well as a host of other brain regions that have the power to stimulate these under some conditions. An example would be the hunger center, which does so when we eat.

Given the fact that innovation is so important to the development of human societies and economies, and that the process of innovation is driven by these motivational brain regions, it's time we ask ourselves: what are these brain regions supposed to do for us? It would be nice to believe that pleasure exists simply to make us feel good – a sweet gift from a beneficent universe. But it's easy to see that it serves a purpose as a primary motivator of behavior. What sorts of behavior were the pleasure centers, pain centers and so forth built to motivate?

The question, "What are pleasure centers *supposed* to do?" is teleological to the core. That is to say, the question asks what is the 'purpose' of certain brain regions. The only fair and complete answer to a question phrased that way, is: "Whatever we choose to do with them, that's what they're supposed to do." But there's another question that I really intend, and it's not so easy to phrase. The real question is: "Given that these brain regions evolved in the heads of my ancestors (a process requiring countless generations of genetic trial and error), and that my ancestors passed the DNA on to me, and that my body grew up as an embryo and then a fetus and then a child assembling the cellular circuitry of those pleasure and pain centers in my own head according to their ancient blueprints... given all that, *why* did my inherited DNA code tell me to build my motivational brain centers the way I built them, and not some other way?"

The reason to ask this long-winded question is that our motivational brain regions must have served *some* function among our ancient ancestors. If we hope to understand our own, modern motivations, then we need to start by knowing why our brains are built the way they are, and not some other way. Each of us built himself or herself, cell by cell, during the process of embryonic and childhood development. We did the work ourselves, under our own power, but the plan for the building process was handed to us by the DNA code we inherited from our parents. That code is the product of evolution. Most of the evolution that created the code was due to the process of natural selection working among our ancestors, both human and pre-human.

Let's start there, with the idea of natural selection. To illustrate how natural selection works, I'll ask you to imagine a green valley where there lives a population of field mice, isolated from the outside world by mountains. These mice spend their days foraging for grass seeds while keeping an eye peeled for hawks. Whenever the breeding season rolls around, the males compete for the attention of females, and the females have their pups, nurse them, and watch over the youngsters until they're big enough to fend for themselves.

The individual mice all look pretty much the same, at least at first glance. But as you walk through this valley, something about these mice piques your curiosity. You soon find yourself spending a lot of time in the valley, picking up mice and having a closer look at them. Pretty soon, you realize that no two of them are alike. Some are a little longer and skinnier than others, some have shorter or longer fur, some are darker or lighter than usual, some have big heads or long tails or broad feet. The mice are as unique as snowflakes, once your eye gets used to looking at them – each is as unique among its own kind as you are among yours.

The endless variation among these mice is partly due to the conditions they've experienced as they were growing up, such as infantile diseases and the availability of nutrition. But a substantial amount of the difference among them is genetic: one mouse inherited unusually large ears from its mother, another inherited a tendency toward obesity from its father, and so on. Like many other mouse species, this one goes through a complete generation every year. As those generations fly by, the genetic traits mix and match among the many family lineages that make up the population. Occasionally, a mouse pup is born with a really bad deal, in terms of the DNA it inherited; for example, its legs might be too short to let it run fast enough to escape hawks. It might even be born with a genetic disease, in the same way that a person can be born with, say, cystic fibrosis. But for the most part, the genetic variation among the mice is harmless, and merely creates interesting diversity among the population, much as it does among us humans.

Your obsession with mouse-watching never goes away, and (since you take good care of your health) the years roll out into decades and centuries, and you find yourself keeping a close eye on these mice for several thousand years. Your devotion to this project is really amazing!

An ice age is coming, and you notice that every winter is getting a little colder than the one before. The change in climate is slow but relentless, adding up to a series of colder and colder years that continues for centuries. Fortunately, the valley is at low enough latitude that it's not going to freeze over completely, which would cause the mice to go extinct. But you can tell that in a few more centuries, by the time the northerly regions of the world are buried under ice, this valley is going to be a tough place for a mouse to live in.

The changing conditions are harder on some mice than others. As it gets colder, any mouse that's born a little fatter, a little rounder, or a little furrier than average has a bit of advantage over the others. Short legs may mean that a mouse is a slow runner, but in cold weather the long-legged mice are prone to frostbitten toes, so short legs become more and more of an advantage. Also, as the winters get longer and snowier, any mouse with pale fur is going to get an advantage over darker mice, because it's harder for a hawk to spot on the snow. Any of these advantages will pay off in the form of enhanced chances of survival, and especially survival to the age of reproduction. Any mouse that dies before reaching that age will not be contributing its DNA to future generations.

In the new, colder conditions, any of the mice that are fat, furry, pale and short-legged are consistently more likely to survive and reproduce. This fact creates a trend that begins to change the population. After a few centuries, you find that you never see any dark, skinny mice any more. It's not because the dark, skinny ones die – mice like that simply aren't being *born* any more. The mouse family lineages that used to carry DNA encoding for dark, skinny pups have suffered so many losses over the years that they've been gradually replaced by those family lineages that are fatter and paler. So not only do the mice *look* different nowadays, the population's DNA has actually shifted to a slightly different code. There's still plenty of individual variation among the mice, both in terms of their appearance and their DNA code, but the average genetic condition has shifted toward fatness and pale fur.

That's evolution. Biological evolution is any directional shift in a population's DNA code. A number of things can cause such a shift to occur, but the most interesting one is natural selection, which is what happened to the mice in the story. If environmental conditions go through a directional change (like getting colder), a population's DNA code can shift to track it. This occurs because those family lineages which carry DNA code that's useful in the new conditions will survive and reproduce more successfully than their neighbors. We call that process 'natural selection' because nature is selecting which lineages persist across the generations, and which ones wither away.

Natural selection is the most interesting cause of evolution because it's the only one that can create innovations in response to environmental stresses (such as ice ages). All other natural causes of evolution are based entirely on random processes. To see how random processes can cause directional shifts, stand a plastic bottle on a small table, and begin shaking the table. The bottle will wander one way and then another, but eventually it will find its way to one side of the table and fall off. The driving force was random, and the path was random, but eventually the bottle made a definitive move in one direction. That sort of process can lead to change, yes, but it can't lead to innovation. You'll never invent something as interesting as a spider web that way.

Because natural selection allows a population to adjust to its changing environment, it causes the population's DNA to shift in a much clearer and more direct fashion than mere random tinkering. Natural selection is 'blind' in the sense that it doesn't look into the future, plan ahead, and make preparations. But natural selection makes up for this blindness by having a very sensitive touch. It detects every little change in the environment and responds quickly (within a few generations), nudging the population in whatever direction causes it to fit better with the changing circumstances. If that directional process continues long enough, it can lead to a lot more than just fatter, paler mice. Through endless tinkering and failed experiments, it can eventually lead to extraordinary innovations – things like the vertebrate eye, the wings of birds and insects, or the human brain. Such innovations start as simple, crude improvements on something that was already present, and gradually elaborate until they become true novelties.

In Chapters One and Two, we considered how innovations accumulate in human societies, gradually increasing the complexity of those societies from hunter-gatherer bands to huge industrial states. Similarly, the novelties created by natural selection can accumulate in some groups of species (including ours) until the organisms in those species are jam-packed with amazing physiological innovations, such as lungs and kidneys. This is the origin of our biological complexity, which could never have been achieved through any of evolution's mechanisms except natural selection.

The driving 'force' behind these directional, evolutionary changes in a population is the ability of one individual to survive and reproduce better than another, under a given set of circumstances. This ability goes under the unfortunate name of *fitness* in modern biology. Few terms in the history of science have created more misconceptions. Worse, the subject matter that we're going to be dealing with – fitness among humans – is the most dangerous place for such misconceptions to arise. We'd better pause for a closer look at the term, to make sure we know where the danger lies.

Although the word 'fitness' is used by evolutionary biologists as a technical term, it arose out of popular, not scientific, usage. The English philosopher Herbert Spencer coined the phrase "survival of the fittest" in 1864, and it captured the imagination of the press and the public. Darwin, in a moment of weakness, adopted the popular phrase and added it to *The Origin of Species* for the fifth edition printing, in 1869. The rest of the scientific community quickly took up the new term.

The problem with the word 'fitness' is that it suggests physical strength and prowess, particularly in those forms which serve in a struggle for power and domination. That's pretty much pure Nietzsche, and stands in direct opposition to Darwin's point of view (and that of modern evolutionary science). The Darwinian idea behind the word 'fitness' is really something like "fitting well into the environment," and the word we really need is something like 'fittedness'... which just doesn't have the same ring to it. But by calling this state of 'fitting well' *fitness*, the generation of scientists after Darwin paved the way for endless misinterpretations based on the pre-Darwinian, Romantic view of "Nature, red in tooth and claw". Most notoriously, the misconception that the only organisms that 'fit well' with their environments were the ones possessed of gritty, warlike 'fitness' justified the appearance of social Darwinism over the following several generations, as described in the Introduction.

Actually, evolutionary fitness can be almost anything. It's rarely a matter of 'physical fitness', strength, or a capacity for asserting dominance. In the valley of the field mice, described above, fitness traits originally included good teeth for chewing grass seeds and strong legs for fleeing hawks. But when the ice age came, new fitness demands appeared, such as being fat and pale. Hundreds of different fitness traits have been documented in thousands of species. Fighting ability *can* be among these traits in some species, sure, but it's nowhere near the top of the list. For one thing, most organisms are bacteria, plants, algae or fungi. Evolutionary fitness is just as crucial among those species as it is among animals, but that quiet majority of life-forms pretty much *never* battle with tooth and claw. Even among animal species, fighting ability and dominance traits are only minor components on the list of real fitness traits.

The descriptive term 'fitness' was redefined as a useful, quantitative measurement by the population geneticist J. B. S. Haldane in the 1920s. Since then, evolutionary biology has been permanently stuck with the term. A lot of modern evolutionary biologists make fitness calculations as part of their routine, day-to-day work, and have a tendency to forget that the word is used in a very different way by everyone else. That can lead to grave misunderstandings, especially if we're discussing the 'fitness' of people.

Let's go back to our mouse anecdote for a moment. Now that we've established that factors like chubbiness and pale fur can perfectly represent the peculiar factor that evolutionary biologists refer to as 'fitness', something else starts to become clear. The evolution of the mice from their original state into the plump, white-furred, snow-adapted mice of the ice age happens because the DNA of the whole population gradually shifts, over the course of many generations. It's not as if some particular mouse, during the course of its lifetime, sweats and strains to become fatter and paler, succeeds in that endeavor, and passes the brilliant innovation on to its offspring. If that were the case, evolution would proceed much faster than it does in real life.

Instead, what happens is something more like this. Let's say a particular female mouse in the valley gives birth to a litter of seven pups. She successfully raises them all, and eventually they leave the den and set off on separate paths into the world. Each of the seven young mice is an individual, slightly different from any of its brothers and sisters. Some are a little fatter or skinnier, some have slightly longer or shorter legs and ears, some have fur that is paler or darker, and so on.

It turns out that this is a very snowy year, during that long progression of years as the ice age sets in. Out of the seven offspring, only two manage to become successful parents. Among the others, the first to go is the unfortunate one with the darkest fur. It stands out so sharply against the snow that it is eaten by a hawk long before the breeding season even rolls around. The other six all survive the winter and make it into the breeding season, but one is a male with funny-looking ears, so no female wants to breed with him. Another settles into a region of the valley that (by sheer misfortune) has a very bad growing season, so she spends her entire life struggling merely to survive. Although her powers of survival turn out to be remarkably good, she never finds enough time and energy to go through the rigors of pregnancy and pup-rearing. Another of the seven siblings turns out to have a reproductive abnormality, so although he settles down and breeds with a female, he sires no pups. The fifth member of the litter is a female that does, technically, succeed in breeding – but she has her litter of pups in a den that's too open to the elements, and they die of exposure before she can wean them. As for the remaining two siblings, everything goes fine for them and their pups, and so the family line is carried forward another generation.

Here's what this story tells us. You don't win the competition of natural selection by surviving, any more than you win it by becoming physically fit and dominating the others around you. If you survive to a ripe old age, putting up a great fight all the while and causing your competitors to bow before you, your evolutionary fitness is still zero if you fail to reproduce. You will contribute absolutely nothing to future generations, genetically speaking. Furthermore, even reproduction is still not quite enough. In order to get DNA into future generations, your offspring must *also* survive to adulthood, and must themselves prove able to carry on the family line.

Now, it's true that natural selection is, irreducibly, a form of competition. But that competition is not for dominance and power, nor even for survival. It's for successful reproduction, not only in the sense of making offspring, but in the broader sense that also includes providing for their welfare. One of the key points of successful evolutionary competition is investing just the right amount of labor and resources into each offspring. In most species, a parent can't afford either to be wasteful, nor to be so stingy that the offspring get off to a bad start. It turns out that the appropriate amount of provision for each offspring can vary over a huge range, depending on the details of the species and its environment. For example, a lot of insects give each embryo just enough yolk to let it survive until the egg hatches, and after that it's on its own. At the other extreme, a human parent might scrimp and save for decades in order to put the kids through college someday.

That observation leads us to one important exception to the simple rules I've been outlining. I said earlier that reproduction is the only way for an organism to get its DNA into future generations... but that's not exactly true. If an individual has brothers or sisters, then they carry copies of a large portion of the individual's DNA code. In fact, if I have a daughter and a sister, then both of them are related to me *equally*, sharing with me the same proportion of my DNA code. Because of this, an individual organism can help perpetuate the survival of its own DNA into future generations without ever breeding, by helping to raise its younger brothers and sisters.

This fact is the basis of an alternate life strategy that's seen in many species – namely, to stay at home and help raise younger siblings. This is called being a nest helper (in birds) or a den helper (in mammals). Consider a species of songbird in which the young fledge into full-grown adults during their first year, and adults continue to lay eggs every breeding season for several years before they die. Caring for the hungry chicks is hard work, and a pair of parent birds may have a lot of trouble getting even one chick to survive to adulthood in any given year. Now, imagine that you are such a songbird, born last spring and now grown to adulthood. Spring is rolling around again, and it will be your first breeding season. You have two choices: go find a mate and struggle to build a nest of your own, or stay at home with your parents, and help them raise a clutch of your little brothers and sisters. Here's a remarkable fact: those two choices work out exactly the same, as far as evolution is concerned. For one thing, as mentioned, a baby sister would carry just as much of your DNA as if you had a daughter of your own. And either way, the eggs will hatch this season – so a new sister gives you just as much hope for the future as would a daughter.

Because of this odd situation, nest and den helpers exist in a lot of species. This is the most common form of what's called 'kin selection': the extension of the idea of evolutionary fitness to include the fitness of close relatives. This turns out to be a pretty powerful evolutionary force, and has led to big, tightly bound extended family groups in many species. But unfortunately, we can't explain the formation of modern human societies in terms of kin selection. There just isn't enough relatedness in big human societies to explain the strong social bonds that hold together a modern nation.

Put it this way: If I reach into the middle of an army of ants and grab two ants at random, I'll always find that they are sisters. Even if there are millions of them, they're *all* sisters. Given that fact, it's not too surprising to see that they'll make great sacrifices for each other. On the other hand, if I pick two random soldiers out of a modern human army, they probably have no kin relation to each other at all. And yet, human soldiers often make great sacrifices for each other, as if they really were brothers and sisters. That in itself shows pretty clearly that kin selection is not a sufficient explanation for the cohesion of our big, complex modern human societies. It explains big, complex ant societies quite well... but not us.

There are a lot of known reasons for animals to live and work together in social groups: they may be using each other to keep an eye out for predators or for food, huddling together for warmth or safety, accessing each other for breeding purposes, and so on. Kin selection in family groups is just one more such reason, but it's an important one, because the individuals in such a kin group are willing to make real sacrifices for each other's benefit. That makes for a strong social group. Some familiar examples of mammal species that live, hunt or forage in kin-based groups are timber wolves, lions, orcas, many types of monkey, several types of mongoose, chimpanzees, and bonobos. The last two examples are of special interest, because they're so closely related to humans.

As described in Chapter Two, there was a time when every human being on earth lived in small, kin-based social groups, called bands. In fact, this was true for the first several thousand generations of our species's existence, and only began to change a few hundred generations ago. The large, complex human societies that most of take for granted – societies such as industrial nations – are a very recent curiosity. If we look back all the way to the origin of our species, we find that in most of the times and places where humans have existed, everyone has lived in kin-based band societies that were held together by the same family bonds that create close-knit social groups in other species.

So *that* is the world we lived in while our brains evolved into their current form. The pleasure and pain circuitry in our brains, which create the basic drives behind our relentless dissatisfaction, evolved by natural selection acting upon our clan groups. For you and me, sitting here puzzling over human nature, that's a bit of good news. After all, it may be difficult to explain our modern human world, but when we look back to our ancestors in their hunter-gatherer bands, it's not hard to see their basic motivations. So let's turn the clock back to 10,000 BC, a time just before the first tentative efforts at agriculture – back when civilization wasn't even on the radar screen yet.

By 10,000 BC, our species, just like every other species of life on earth, had undergone endless generations of fine-tuning through natural selection. Circumstances had mercilessly trimmed away any family lineages among our ancestral populations that consistently failed to get DNA into future generations. What remained after all those millennia of cruel trimming was a human world of extended-family groups, each an independent society – a sort of micro-nation. Each of these bands consisted of a handful of people adapted and acculturated to compete against their neighbors in the only game that mattered: raising the next generation. It had been that way since before our species even existed. By 10,000 BC, the winners of that game had been inheriting the earth, generation after generation, since long before anyone ever stood on two legs.

The motivational circuitry that's found in our brains today arose during that long, pre-civilized period, including the many generations before we became *Homo sapiens*. By 10,000 BC, our brains were physically identical to the ones inside our heads right now. The circuits that control our behavior and thoughts today were designed by natural selection operating on hunter-gatherer bands *before* the innovations of agriculture and of cities. It's true that we now use our brains for different things than our ancient ancestors did, but the core circuitry is still there, hardwired into our heads, just the way we inherited it. We can't change the circuitry of our brains any more than we can sprout wings.

Figure 4. The Darwinian purpose of life.

None of us would be here today if our ancestors hadn't succeeded in perpetuating their DNA into future generations. Their success in achieving that goal was measured only in terms of the number of children they raised to adulthood – either their own children, or those of very close relatives.

How much does it matter that our brain circuitry originated in such ancient and unfamiliar circumstances? In some ways, it doesn't matter much. We can modify our own behaviors whenever we want to work around our inherited motivational tendencies, and that ability lets us do pretty much anything we set our minds to. Maybe we can't sprout wings, but we can build airplanes. But here's what we *can't* do: we can't make ourselves really want things that we don't want. To some degree, our desires and specific dissatisfactions, our core drives, are simply built into our brains. As this book continues, we're going to see that that turns out to matter a great deal.

So the roots of our modern dissatisfactions are found in brain structures that evolved to let human bands, which is to say extended family groups, get their DNA into future generations. The task those bands were facing seems straightforward enough, in principle. Babies are the key asset that must be produced in order to win the Darwinian competition for inter-generational survival, so the primary challenge was to make babies. No problem there: humans are often willing to labor diligently on the project of conceiving a baby. The main things that limited a hunter-gatherer band's supply of new babies were the social constraints within the clan and the physical constraints of reproductive biology (which in turn were influenced by food supply and other resource issues).

But there was more to it than that. Producing a baby is only the very beginning of the reproductive process. To truly reproduce means to create a new generation, one that carries your DNA into the future. That generation isn't fully formed until your babies have successfully grown to reproductive adulthood. Only then have they "filled your shoes," so to speak. That part of the reproductive process, the part that comes *after* birth is achieved, turns out to be an immense investment for humans. In fact, it's one of the most conspicuous ways in which our species is different from the rest of the animal kingdom. Humans take longer to raise our offspring to independent adulthood than any other species on earth.

Back when we lived in hunter-gatherer bands, the survival of children to reproductive adulthood depended on a combination of luck (due to childhood diseases, etc.) and the investment of care (providing food, protection, survival training, etc.) A great deal of the investment in childcare was indirect, and wouldn't have looked like childcare at first glance. For example, imagine a Paleolithic father who spent weeks at a time out in the forest with his buddies hunting game and fighting rival bands that tried to push across his band's territorial borders. It might seem like this man was leaving the care of the children entirely to his long-suffering wife back in the home village. But from a Darwinian view, we're likely to find that his actions were perfectly optimized to maximize the welfare of his children. Even if this man never thought for a moment about anything except his own pride in his strength of arms, we can probably look back and confidently say that he never did a single action of lasting meaning in his whole life except putting food in his children's mouths. His DNA lived on after he died, in the form of his children; everything else that he did and said and represented was buried along with his body. It's quite likely that his DNA is still with us today, but the rest of him is utterly forgotten.

Pretty much everything that happens in a hunter-gatherer band can best be understood as an effort toward successful reproduction, however indirectly the effort might be applied. Any other motivational tendencies are constantly being weeded out by natural selection. But in order for a clan group to maximize its chances of persisting into the next generation, its limited resources must be doled out carefully. In particular, a hunter-gatherer band must make wise trade-offs between the resources spent on having new babies and resources spent on giving care to the existing children. Such decisions require long-range planning of a sort that most animal species never have to deal with, because their offspring don't require as much care as human children. A songbird might lay eggs in early April, and by late May the chicks are fully grown and fly away. If you're a human parent, you're not going to get off that easy.

Putting these observations together, we can make a general description of certain essential aspects of the lives of our ancestors in 10,000 BC, which was about 600 generations ago. First, we can say that the clans (or band societies) that existed at that time were those whose ancestors had been managing to get DNA into the next generation, one generation after another. We can say that the set of requirements for getting that job done in those human societies were very similar to the requirements imposed upon lots of other species, especially other species living in extended family groups, such as wolves and chimpanzees. We can say that these conditions had existed for a very long time. In fact, similar social conditions had been in place not only since our origin as a species but beyond that, back to the species that preceded ours, and the one before that and so on, probably going back several million years.

That doesn't mean that these Darwinian conditions still apply to us, in our modern civilizations. In fact, they don't. The reason for that is that we have largely broken free from the constraints of natural selection, as we'll discuss shortly. Natural selection is a widespread principle of biological evolution, but it can become negligibly weak in certain circumstances, and our modern societies turn out to be living under such circumstances. But before we get into all that, let's finish looking at those tiny, clan-based societies that were the social structure of our lineage for millions of years, even before we became human. It was during that span of time that our brains gradually evolved into their modern form, tripling in size as they did so.

Exactly what sort of evolutionary pressure was at work during those millions of years while nature was designing our brains? By now it's clear that it would be hopelessly naive to suggest that some sort of underlying drive toward power and dominance was at work, or even an underlying drive toward complex behavior and intelligence. Evolution cannot intrinsically favor a will to power, or a trend toward complexity. The only pressure that drives evolution is the one that enhances a given lineage's DNA by increasing its number of surviving offspring. Of course, strength and intelligence can both play roles in enhancing that number under some circumstances, just as fatness and paleness were key factors in our mouse example. Under other circumstances, natural selection can place equal favor upon physical weakness and stupidity. In fact, the latter sorts of pressure are vastly more common in real life, and that's why so many highly successful species are small, weak and stupid (for example, various cockroaches and bacteria). What they are good at is reproducing, and that's the trump card.

Now we're ready to answer this chapter's opening question: Where did our brains get their pleasure centers? We've already seen that the parts of the brain that give us our essential motivation (our feelings of dissatisfaction) evolved by natural selection. That evolutionary selection happened during the long pre-civilized period of our ancestors' existence, a period extending back countless generations, even before we became *Homo sapiens*. During the long millions of years when we lived in small hunter-gatherer bands, we were exposed to plenty of natural selection, and that's what expanded the one-pound brains of our ancestors, four million years ago, into the three-pound version we've got today.

Our brains grew larger and more innovative as those years passed, but only as a tool to support the constant, unceasing quest to maximize the amount of DNA each clan passed on to the next generation. The reward systems in our modern brains exist to encourage behaviors that – directly or indirectly – caused our ancestors to live lives that included having children and caring for them. Those behaviors include many examples of physical prowess and aggression (such as skills of warfare and hunting) and many examples of intelligence and imagination (such as skills of innovation). But evolution only worked to enhance these qualities in our ancestors to the degree that they contributed to each clan's ability to produce children and to care for them.

Given the pressures of life and death in ancient societies, the driving power of these child-making and childrearing motivational centers in the human brain must be pretty strong. It's time, then, to have a look at the kit of temptations and threats that is concealed inside our heads. What are these motivational circuits that have bossed us around for all these millions of years, trying to coax us into having kids?

In order to create a new generation, humans have always had need of a drive that leads to the conception of embryos. But, incredibly enough, we don't actually *have* any drive toward conception – not as such. In us, as in most of the animal kingdom, the drive that stands in for that missing motivational system is the sex drive, or libido.

Humans, like other mammals, tend to transfer sperm cells to within swimming distance of the egg during the course of carrying out behaviors that are driven by the libido. A human couple living in Paleolithic times might hope to conceive an embryo or they might not, but as long as they keep carrying out the behaviors driven by their libidos, they're likely to conceive one sooner or later. Even if embryonic conception is the *last* thing on people's minds when they are obeying their sex drive (or for that matter, even if it seems like a catastrophe), still the libido is a very effective replacement for the missing drive toward conception. Our brains may not contain any drive to build embryos, but we have a mighty powerful drive to carry out activities that tend to create them as a side effect.

In terms of brain circuitry, the libido is one of the most basic physiological drives. It's regulated from the hypothalamus, much like hunger, sleepiness and other simple drives. The libido gets its motivational power over other parts of the brain by using the dopamine reward system centered in the nucleus accumbens. You'll recall that this is the way the hunger drive works, as we saw in Chapter Three.

Once a woman in an ancient hunter-gatherer band was pregnant, the only major behavioral change that was necessary to support the prenatal nine months of development was that she must eat more food. That's directly mediated through her brain's hunger centers. But once the baby was born, in addition to the agonies of childbirth, the woman (along with her husband, mother, sisters or whoever in the clan was supporting her efforts) would find herself saddled with an immensely increased workload. These postnatal costs were so conspicuous that a woman might be expected to avoid a second pregnancy by every possible means – libido or no libido. One way in which the human brain discourages that form of acquired wisdom is by providing direct, reward-based motivation systems for pregnant women and new mothers.

The first of these is a direct chemical link between the embryo and the brain of its mother, via the mother's bloodstream. Embryos release beta endorphin into the mother's blood, and this chemical has effects very similar to heroin. It creates a euphoric, abstracted state of mind, and helps make pain easier to bear. It's also highly addictive. The exact function of the embryo's release of beta endorphin is still speculative, but an obvious interpretation is that it reduces the suffering of pregnancy and childbirth, thus neurologically rewarding the mother for her behavior of getting pregnant. As with opiate drugs like heroin, the most common side effect is nausea, which manifests as 'morning sickness'. Opiate drugs also cause withdrawal symptoms when the supply is removed, and some authors believe this is the basis of post-partum depression. At any rate, some women say they enjoy pregnancy and some do not, so the efficacy of this reward system (if that's what it is) is presumably hit-or-miss.

The most important set of brain circuits, other than the libido, that serve as a direct drive toward making and raising children was discovered just a few years ago, though I think everyone expected such circuits existed. A 2009 study used functional MRI, which is a brain-scanning technique, to observe brain activity in childless women as they looked at pictures of babies. The sight of a cute baby stimulated activity in the pleasure center, the nucleus accumbens. There's no reason to imagine that childless women are the only people who have this reaction to cute babies – it's quite possible that everyone does. The brain is still largely a mysterious organ to science, but we can now say with confidence that there is at least one core motivational system that's not merely intended to cause humans to make babies, but also to cause us to hold, protect and care for them.

So our brains, as handed down to us by our ancestors, contain neurological circuitry that's directly oriented toward making us enjoy the production and raising of children. But that's only a small part of what's found inside our skulls. Each of us built his or her own brain, cell by cell, during the years when we were embryos and children, and we did the whole job using a DNA blueprint that was drawn up by the forces of natural selection, long before we were born. Those forces of natural selection were only influenced by one form of success or failure, namely the degree of success in placing DNA into future generations. That's where your brain's blueprint came from. Regardless of how we think of our lives, or how we choose to live them, *every* circuit and region of the brain evolved as a hardwired structure intended ultimately to facilitate the production and raising of children – either your own children, or those of very close relatives.

The brain regions that directly goad us to have children and care for them are just the tip of that iceberg. The main bulk of the brain works toward the same end, but through a huge variety of indirect means. These include desires to do various activities that, long ago, indirectly supported reproductive success in our ancestors. They also include our most sophisticated human capacities, such as the ability to plan and execute a disciplined sequence of actions that requires strategic suffering in order to improve a situation.

Here's a simple example, to show how natural selection can force the evolution of brain circuits that indirectly increase a person's number of surviving offspring. Let's imagine that I'm living in a hunter-gatherer band in the year 10,000 BC, before agriculture and civilization existed. It's late spring, and the first berries are ripening on the shrubs around my village. I put a ripe berry into my mouth. Its sugars stimulate the sweet-detecting taste buds on my tongue. The sensation is carried from tongue to brain through my seventh cranial nerve. Inside my brain, the signal is sent to the anterior insula of the cerebral cortex, and from there to several regions for analysis. One crucial pathway is a signal to the nucleus accumbens, where it causes release of dopamine. What that means in terms of my personal experience is that eating the berry gives me a little tweak of pleasure.

That reliable burst of pleasure is likely to reinforce my tendency to eat berries. Of course, I'm not a machine, and I can override that tendency as part of a teleological plan. For example (since this is 10,000 BC), I might harvest a lot of berries and trade them to someone for a new arrowhead. Either way, whether I eat them or trade them, the berries have slightly decreased my chances of starving to death in the near future. Seen in the long term, that's going to give me slightly better odds of leaving behind some DNA after I die. Naturally, I'm not thinking about any of that while I'm gobbling the fresh berries. But I don't have to: Those of my generation whose brains give out dopamine rewards for berry-eating are going to pass along a little more DNA than everyone else, not because we're smart but because we're well fed. The result? By the 21st century, pretty much every human on earth likes the sweet flavor of berries.

Pain circuits evolved by a similar mechanism. Once again the real issue, the thing that drove evolution, was not the painful event itself, but rather the number of grown-up children the pain sufferer left behind when he or she died. Let me again put myself into the role of an ancestor from 10,000 BC. Living in that wild and ancient time, I try to avoid seriously injuring myself, because whenever I do so it hurts like crazy. The neurological circuits (centered in my dorsal posterior insula) that give me this internal 'punishment' are strong motivators, discouraging me from doing dangerous, self-injurious actions. A few of the other people around me don't seem to care as much as I do about getting hurt, and they're more prone to getting nasty wounds. Their wounds often lead to infections and an early death. As a result, those careless individuals, with their poorly formed pain circuitry, are less likely to survive long enough to have and raise children. The long-term result is that I pass on more DNA than they do, and thus, by the 21st century AD, pretty much everyone has brain circuits that make them dislike pain.

So, our brains still contain circuitry that evolved under the conditions of natural selection, a force that put constant pressure on our ancestors, long ago. This pressure worked to cause each family to produce as many healthy adult offspring as circumstances would allow, generation after generation. Regardless of whatever other virtues and strengths a family might have possessed, if those strengths didn't lead to a boost in the number of healthy adult offspring, then other families gradually replaced them as the generations went by.

When you think about it, that's an odd place for our modern brains to have undergone their design phase. Hardly any of our behaviors nowadays are the same as they were in 10,000 BC. So then... what does natural selection have to do with *us*? There's a question that's going to open up a great big can of worms.

6

GETTING FREE FROM NATURAL SELECTION

It's time for me to admit to the reader that modern, civilized human social behavior doesn't really fit very well with simple evolutionary descriptions. I've been avoiding talking about civilized societies by keeping our attention focused on the conditions in which our neurological reward systems *first evolved*. But now it's time to ask: Why are modern civilizations so different from every other social system in the natural world? We're conspicuously different both from early humans, and also from every non-human social species. Not surprisingly, the answer to this puzzling question becomes clearer as soon as we realize what's unusual about the way we pass our DNA on to future generations.

To get your DNA represented in future generations, you must not only survive and reproduce, it's necessary that your offspring grow up and reproduce, too. That's not just a human issue; it's universal to all life. Each species, including ours, must optimize the way it splits up its limited resources between the project of *making* offspring and the project of *raising* offspring. Our ancient ancestors struggled to get that balance just right, as most species do, and it was in that context that our current brain structure evolved. This point is important enough that I'm going to illustrate it with a brief story about three species of animal.

Our first species of imaginary animal offers no care to its offspring. It simply lays eggs in a protected place and runs off, hoping that no predators have seen where the eggs were laid. When the young are born, they're on their own, and the great majority of them naturally get picked off by predators before they can grow up. During adult life, members of this species compete for space and food, and when the mating season comes they compete also for mates. But competition or no, only one factor really matters as the generations go by: How many eggs did a given individual lay? The eggs are like lottery tickets. Most of them will be of no value in terms of contributing to the next generation, but a tiny percentage of them will get lucky, survive the horrors of youth, and grow up to become full-fledged adult members of the next generation. The greater the number of eggs a given individual lays, the more lottery tickets she has. If all else is equal, the number of eggs laid will be the sole key to getting DNA into future generations.

For the second species, circumstances are a little different, and the adults find they can improve their chances of winning the reproduction lottery by putting extra care into individual offspring. They guard their eggs, feed their babies, and when the young are big enough to get around, the parents take the offspring with them on foraging trips and teach them how to find food and stay safe. The cost of all that investment is that the parents can't care for more than a few offspring at a time... let's say three or four per breeding season, in this particular species.

Evolutionary 'fitness' in this second species is a little more complicated than an egg lottery. If two competing breeding pairs have equal access to resources, and one of the pairs produces three babies while the other produces four, the pair that produced more offspring does *not* have an indisputable advantage. The four-baby parents are going to have to split their limited resources four ways among their young, while the three-baby parents can put 33% more investment into each offspring. If that extra investment of time and materials amounts to a big enough advantage for the offspring, it could easily mean that the three-baby family has a higher chance of seeing one or two offspring grow into reproducing adults. If that happens, strange to say, their route to evolutionary victory is to have *fewer* babies.

Now picture a third species, again one that must place a huge investment into each offspring in order to keep it alive until adulthood, just like the second species. There's some threshold quantity of time and material which must be invested if a breeding pair wants to be (say) 85% certain that a given newborn will live to adulthood, and then through its reproductive years. We'll call this threshold quantity of investment: X. This quantity, X, is a whole lot of resources, so this species has evolved to produce babies slowly... let's say that each breeding pair produces just one per year, at best. Things have been this way for so many generations that nowadays their bodies are simply built accordingly, so one baby per year is the fastest rate that their reproductive systems will allow.

Now imagine that we intervene and artificially enrich the environment of this third species. We keep saturating the region with food and other useful materials until *every* breeding pair of the species can obtain X resources every single year – not only that, they can actually get 10X or even 100X every year! Under these oddly rich circumstances, the pressures of natural selection go back to the simple conditions that we saw in the first case, the 'egg lottery': each family lineage will get its DNA represented in future generations (its evolutionary fitness) by a simple 'numbers game', and the only number that matters is how many babies each breeding pair can produce. Issues regarding the *quality* of care no longer affect the outcome, in terms of how much DNA gets transmitted to future generations. If one breeding pair invests 100X into each offspring, while their neighbor only invests 3X into theirs, still both families' young are almost certain to grow up and join the breeding population. The extra 97X investment has no effect, in terms of natural selection.

Under this weird set of circumstances, the term 'evolutionary fitness' can simply be replaced with the phrase, "the tendency to have children." All other 'fitness' traits (successful competition, being fat and pale, fighting ability, innovative intelligence, whatever) all fade into insignificance, and only one factor has any substantial influence on the amount of DNA that a family lineage contributes to future generations. That factor is the number of children produced by the family's females of reproductive age.

What is this strange, highly manipulated third species? Well, that's us. The third case that I've described in this story is precisely the case of human beings in the wealthy industrial nations of the modern world.

That story's punch line implies that I believe that modern humans often live with a super-abundance of the resources needed for having and raising children, in terms of ensuring that they survive from conception through their own reproductive years. It's not hard to back up that assertion. Basic resources are so abundant in the highly developed nations that a marginal increase – or even a doubling – of prosperity has practically no influence on how many babies a family can succeed in raising to adulthood, if the family is inclined to do so.

My story about the three reproducing species also suggested that there's a level of prosperity, or resource availability, that's so high that any further increase has almost no effect on natural selection. Let's give this level of resource abundance a name, and call it the **post-Darwinian threshold**. In terms of human economies (where we use the word 'wealth' to mean access to abundant resources), the post-Darwinian threshold is the level of prosperity beyond which increasing wealth does almost nothing to affect survivorship of offspring to reproductive adulthood. Can we calculate the post-Darwinian threshold for humans? We certainly can. In fact, it turns out to be remarkably easy to do so.

The average newborn girl in a nation where per capita GDP is about $5000 (US 2015 equivalent value) has an 85% chance of living to age forty, which will get her through her main childbearing years. That level of prosperity is currently found in Pakistan, Chile and Hungary, to give a few examples. I'll spare the reader a tour through the numbers here, but if you'd like a closer look, please flip to Appendix Two, Topic B.

I'm not saying that living on $5000 a year makes people happy or comfortable, or gives them a standard of living that I would personally consider decent or acceptable. I'm merely making a mathematical observation: we can correlate a certain economic condition (the equivalent of living in the US on $5000 a year in 2015) to a certain population-genetic condition, namely a near-optimal opportunity for passing DNA to future generations. Below that income level, humans are severely exposed to natural selection in the competition for resources, just like every other species. Above that income level, we are not.

Now, when I say that humans in wealthy societies are not exposed to natural selection, I'm only speaking statistically. Obviously, a child who finds that there's nothing in life that's more fun than running across busy interstate highways is unlikely to grow up and communicate his DNA into future generations. Modern humans aren't magically immune to all the usual pitfalls of natural selection – disease, violence, reproductive problems, etc. But as prosperity rises, it does shield us from these things, bit by bit. The average citizen in a rich society has only a marginal chance of suffering the ultimate Darwinian blow, namely the arrival of death (or permanent sterility) before the age of reproductive maturity. The chance that I, personally, am going to starve to death this year is an awful lot smaller than that chance was for any given ancestor of mine, a hundred generations ago.

A long time ago, the societies that are today's developed nations presented their average citizen with a much harsher existence than we see around us, today. So we might ask: How long has it been since typical people living in the cities of wealthy nations were strongly exposed to natural selection? Remarkably, the answer is just fourteen generations or fewer. Holland was the first nation to achieve a per capita level of prosperity that was equal to the post-Darwinian threshold, and they did so around the year 1700. Prior to that, no society in history was ever that rich on a per capita basis, not even imperial Rome (though, of course, some individuals and classes achieved vast wealth even in very ancient societies). The average person on earth crossed the threshold around 1945. Today, the average human being in our world makes nearly triple that amount.

A typical person whose income level is above the post-Darwinian threshold has a potential level of evolutionary fitness that's nearly at the natural maximum for the human species. If that person *wants* to raise a huge family, there's really nothing to stop him or her from having one. Any further addition of prosperity to that person's life may add comfort or even happiness, but it can't possibly give more than a tiny, marginal boost to fitness. Almost all of that person's children would have survived to adulthood anyway, even without the boost in income. If a society's prosperity level is already three times the post-Darwinian threshold, and then doubles again, the improvement in the survivorship of children is usually just a few percent. If we're calculating a family's evolutionary 'fitness', then that tiny boost in survivorship will be utterly swamped by more powerful factors, such as personal decisions about family size.

So, in any modern nation where the average person is making more than $5000 a year, the 'Darwinian fitness' of a person or family lineage consists almost entirely of one factor: the voluntary tendency to have children. How those children are raised becomes almost completely irrelevant to the family's fitness, as are any considerations of the family's social status, physical strength, IQ, wealth, etc. At income levels higher than the post-Darwinian threshold, none of those factors makes more than a tiny difference in terms of how much DNA the family passes on to future generations (and remember, that's what evolutionary fitness is, by definition). Most especially, the tiny differences that *are* created by those factors pale into utter insignificance when compared with the big one: the fact that some individuals want a lot of kids, others want just a few, and some want none at all.

Since this book is really about modern, human societies, we've just stumbled across a great piece of luck. From here on, we can replace the misleading technical term 'fitness' with the phrase "the tendency to have children" whenever we're talking about people who live in the developed nations of the modern world. For the residents of all but the poorest modern nations, the two terms are effectively synonyms. This is a lucky break, because almost nothing gets people more confused and argumentative than talking about the evolutionary fitness of some group of modern humans. We've just found a way to drop the term entirely from such discussions.

So we modern people find ourselves effectively free of natural selection, and may choose either to pass our DNA along to future generations or not, as a simple matter of individual preference. What effect does this freedom from natural selection have upon our evolutionary process? Without the specter of natural selection constantly looming over us, threatening to kill us at a moment's notice, there is a great relaxing of the forces that have guided our evolution as a species. Consider again the example I gave about the evolution of our enjoyment of berries. People evolved brain circuits that made us love to eat berries because eating them added some much-needed calories to our diets in ancient times. The berry-loving circuits are still there inside our brains, but the pressure to get calories from seasonal sources of wild berries is a thing of the past. We can get all the calories we need at the local grocery store. If a new genetic strain of people arises who lack the physical ability to enjoy the flavor of berries, they'll do just fine. Natural selection will do nothing to eliminate their DNA from the human gene pool, nor should it do so. Nowadays, it doesn't matter whether you like berries or hate them. We can eat almost anything we please, and our kids will still grow up and have kids of their own – or not have them, whichever they prefer.

It's hard to complain about having won such luxurious freedom, after all those millennia that our ancestors endured under the cruel whip of natural selection! But perhaps there's a slight feeling of vertigo as we realize how free we really are. Natural selection was an unforgiving master, but it *did* guide us through all the major changes that we passed through during the course of a highly eventful 3.4 billion years. Natural selection took us all the way from being bacteria to being civilized people. Our human populations only began winning freedom from natural selection about fourteen generations ago. What's going to guide us, now?

The answer is that we will guide ourselves, by doing whatever we want – and we will want whatever the reward circuits in our brains *tell* us to want. Sure, natural selection built those reward circuits, and fine-tuned them to get certain jobs done. But with natural selection out of the picture, we can seek to reduce our feelings of dissatisfaction any way we please.

That observation is not something that the reader should accept without thinking it over. The reward circuits in our brains are hardwired, so to speak – we certainly can't change the physical layout of those circuits by simply wanting to do so. No behavior short of radical neurosurgery is going to move the connections around in your brain. It took millions of years for evolution to place those circuits into the exquisitely detailed pattern that creates our human consciousness and intelligence. So, is it really possible to redirect our brains' reward pathways away from their original, evolved purposes?

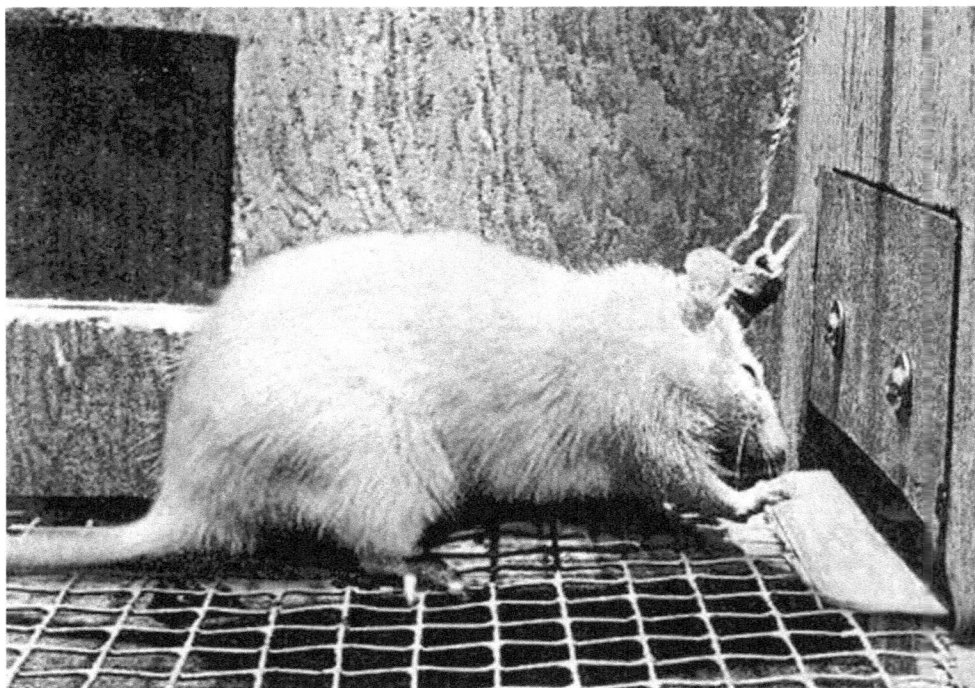

Figure 5. Rat in ecstasy.

This rat has a tiny electrode planted in its brain, positioned where it can stimulate the nucleus accumbens. The rat is pressing a lever that turns on an electric current for a moment, causing a tiny burst of dopamine to be released from its pleasure center.

Indeed it is. Nothing could be easier. The most straightforward way to go about it is, in fact, to do it surgically. Figure 5 depicts one of the first experiments showing the existence of a pleasure center, in the region now recognized as the nucleus accumbens. Experimenters inserted tiny electrodes into the brains of living rats. The rats were given access to a lever that sent a quick pulse of mild electricity through the electrodes, releasing a brief dopamine 'reward' from the pleasure center. The rats were seen to press the levers over and over, a couple of times a second. They would sometimes do this behavior for over 24 hours at a time without pausing to rest or eat.

A more familiar example, and one that doesn't involve any surgery, is the use of pleasure-producing drugs. Consider cocaine, which is a drug that enhances the effects of the dopamine released by the nucleus accumbens. It's easy to addict a rat to cocaine in the laboratory, and cocaine-addicted rats behave much the same way as rats with electrodes stimulating their pleasure centers. We humans, too, can and do become addicted to cocaine, so this is no idle observation. It's an all-too-familiar aspect of modern life: any drug that directly stimulates the brain's reward circuits is likely to shift the user's behavior dramatically and chronically. The adaptive, fitness-enhancing behaviors that the reward circuits originally evolved to stimulate become obsolete, and the person shifts to behaviors that are more efficient at producing the reward. Pretty often, that involves moving into a marginalized social role that may be characterized by crime, poverty and failing health.

Drug addiction gives us a clear example of the brain's reward pathways being redirected away from their evolved purposes. Still, it's an extreme case. Most of us aren't addicted to cocaine or any other euphoric drugs. But now that we're free of natural selection's stern and merciless guiding hand, we're also free to shift our brains' reward circuits away from their original, evolved goals in lots of other ways. Such shifts are usually less dramatic than drug addiction, but they are vastly more common and insidious. Our lives are full of them, and we indulge in them constantly.

The very first place to look for these shortcuts to satisfaction is among the embedded innovations that surround us. In fact, refined cocaine *is* one of those innovations, even if it's an extreme example. In Chapter Four, we saw that innovations are always intended to provide more effective means for decreasing individual dissatisfaction, either directly or indirectly. That means that every human innovation acts to bypass or trick some circuit in the brain that originally evolved to encourage behaviors that first arose through natural selection. Innovations give us more efficient routes – shortcuts – for satisfying our brains' motivational centers. We still have the same neurological circuits in our heads that we had in the year 10,000 BC, but we no longer have to hunt bison with stone-tipped tools to get our dopamine rewards. Our core drives are still the same, but the huge hoard of innovations that surrounds us allows us to satisfy those drives much more efficiently.

In almost every way, these innovation-based shortcuts are a huge improvement for us. Our behaviors, as a species, have changed radically and thoroughly in the past few hundred generations, but not many of us would want to go back where we came from. The conditions endured by our ancient ancestors were unimaginably unpleasant. The circumstances of our modern, civilized societies are hugely improved, and so a lot of the behaviors that our brains evolved to reinforce are obsolete now. I still have the physiological ability to produce a mighty burst of adrenaline if I'm ever attacked by a hungry, carnivorous beast, and that adrenaline would help me to run like crazy toward the nearest tree. But I've never actually encountered that situation during my sheltered life, here in modern civilization.

Not only are we more comfortable nowadays, living in our synthetic modern environment with its huge trove of embedded innovations, but furthermore our opportunities for evolutionary fitness are higher than our ancestors ever dreamed of. As we've seen, the entire basis of natural selection, which is evolutionary fitness, has boiled down to a single factor in modern, developed nations: each individual's voluntary tendency to have children. As long as the new, alternative paths to satisfaction offered by innovations don't interfere with our tendency to have children, they offer no threat to the persistence of our species. Just the opposite; they give us greater opportunities for fitness than ever seen among humans before.

But some innovations *can* erode our tendency to have children. They do this by providing more effective ways to satisfy the pleasure-giving circuits in our brains, which originally evolved to reward us for actions that led to conception and childcare. That may not sound like such a big problem, but it turns out to carry more weight than appearances might suggest. We humans perform our voluntary actions in order to decrease our dissatisfaction. If that dissatisfaction comes from a need to conceive and raise children, and we can find more efficient ways to squelch that dissatisfaction, then we'll never bother to have kids.

This is a very new sort of social issue. Its effects upon early societies, or even those of a few generations back, were downright negligible. To see why that's the case, we'll need to take another quick glance back at the agricultural societies that led to the first civilizations.

Early in the development of any agricultural society, only a relatively small number of innovations have accumulated. If I imagine living in one of these societies, I find that life is still enough of a struggle that the satisfaction of my individual desires is closely tied to my fundamental, evolved needs. Anything that makes me feel less dissatisfied is also likely to enhance my fecundity. In other words, it's likely to increase my reproduction-increasing behaviors, such as those behaviors that promise to keep me alive long enough that I can have some children and raise them. Living in a society that's at such an early stage of economic development, the main sources of my personal dissatisfaction are things like periodic and chronic hunger, and perennial damage incurred by raids from outsiders, and epidemics caused by squalid social conditions. Any innovation that fulfills my constant longings to ease these troubles is also likely to increase my tendency to survive, reproduce, and raise children. In other words, in a pre-civilized society, terms like 'social advancement' and 'economic development' are practically synonymous with 'Darwinian fitness'.

But this fine, clear link between gratification of personal desire and the tendency to raise children becomes progressively weaker once a *lot* of innovations have accumulated. True, ancient people who lived with only basic innovations at their disposal received clear Darwinian benefits from those innovations, such as extra food to give their children. But at the opposite extreme we can find wealthy, modern people who have such a rich and effective kit of innovations that they are able to endlessly carry out behavior patterns that delight them, while never delivering any DNA into future generations. Like a rat with a wired head, it's possible for such a person to stimulate reward circuits in his or her brain without bothering to go through the laborious process of raising children, even though that's what those neurological reward circuits evolved to encourage. Most of us live somewhere in between these two extremes, but the constant buildup of innovations and wealth in our civilized societies keeps pushing us further from the first condition and closer to the second one. We're getting more post-Darwinian all the time.

Artificial, innovative methods of decreasing our dissatisfaction compete against the original, evolved methods of doing so. Over and over, innovative new mechanisms and methods replace the ancient ones, giving us easier routes toward satisfaction of the desires built into the circuitry of our brains. If you'd like to see one abundant set of examples, just consider what we eat. Hardly any of the foods that we modern people put into our mouths can be found in nature. Almost all the plants and animals that are the primary sources of our foods are genetic novelties created by human breeding schemes – and many of these schemes have been in play for millennia. In a number of cases, such as that of the cow, the naturally occurring species that led to the domesticated ones don't even *exist* anymore. The foods generated from these altered species are almost always processed by grinding, mixing, cooking and so forth, often yielding a single foodstuff that's made from dozens of separate species of plant and animal, as well as mineral sources. Nature does not provide us with anything that resembles a pizza, or a Peking duck.

The reason we eat processed foods such as boiled grains and grilled meats, rather than the raw game and roots that sustained our most ancient ancestors, is because these innovations work better to decrease our dissatisfaction. We like their flavors more, and find them less likely to cause sickness or indigestion. The neurological circuits that cause us to like one food rather than another may have evolved before anyone invented cookies, but cookies nonetheless turn out to be a very effective way to satisfy those circuits. Our innovative foods, and for that matter all of our innovations, compete against the older, pre-innovation methods of attaining satisfaction.

That competition can be very direct, sometimes. This is hardly surprising, since the 'natural' or original methods of satisfying our brain's basic drives all evolved under very different (and very harsh) conditions, long ago. We can use the term **short-circuited drive** to describe a motivational circuit in a person's brain which the person satisfies through innovative techniques, rather than going through the behaviors that the circuit originally evolved to encourage. As a simple example of short circuiting a motivational drive, consider the act of grabbing some fast food when you're hungry. The hunger circuitry of our brains once motivated people to go through elaborate patterns of behavior in order to get food, such as chipping a sharp edge onto a stone to make a hunting weapon. When you stop off for a quick snack from a drive-through window, you've just short circuited the drive that led to all of that laborious behavior in ancient times.

As we saw earlier, another and more extreme example would be smoking crack cocaine. That single, easy action can allow you to short circuit pretty much *all* of your evolved motivational circuits in one go, along with all the dissatisfactions they create. This begs the question: If civilization is a pile of innovations, each of which is a gimmick to short circuit our neurological drives, then why aren't we all a bunch of crack addicts? This is an important question, and not nearly as flippant as it might seem. Remember, quite a large number of people in modern societies are, indeed, addicted to dopamine-enhancing drugs such as cocaine. That indicates that the sociological pressure to exploit this ultimate shortcut to satisfaction is quite real, even if conventions make such a lifestyle unthinkable for most people.

To understand why most of us aren't drug addicts, despite the wide availability and impressive effectiveness of pleasurable drugs, remember the Story of the Three Brewers, in Chapter Four. The First Brewer in that story was an alcoholic who enjoyed himself for a short period, then lost everything because of his inattentiveness to his daily business. Even if we imagine that he was living in a rich society that was, overall, beyond the post-Darwinian threshold, he could still succeed in destroying himself and his evolutionary fitness that way, if he just drank hard enough. *Some*

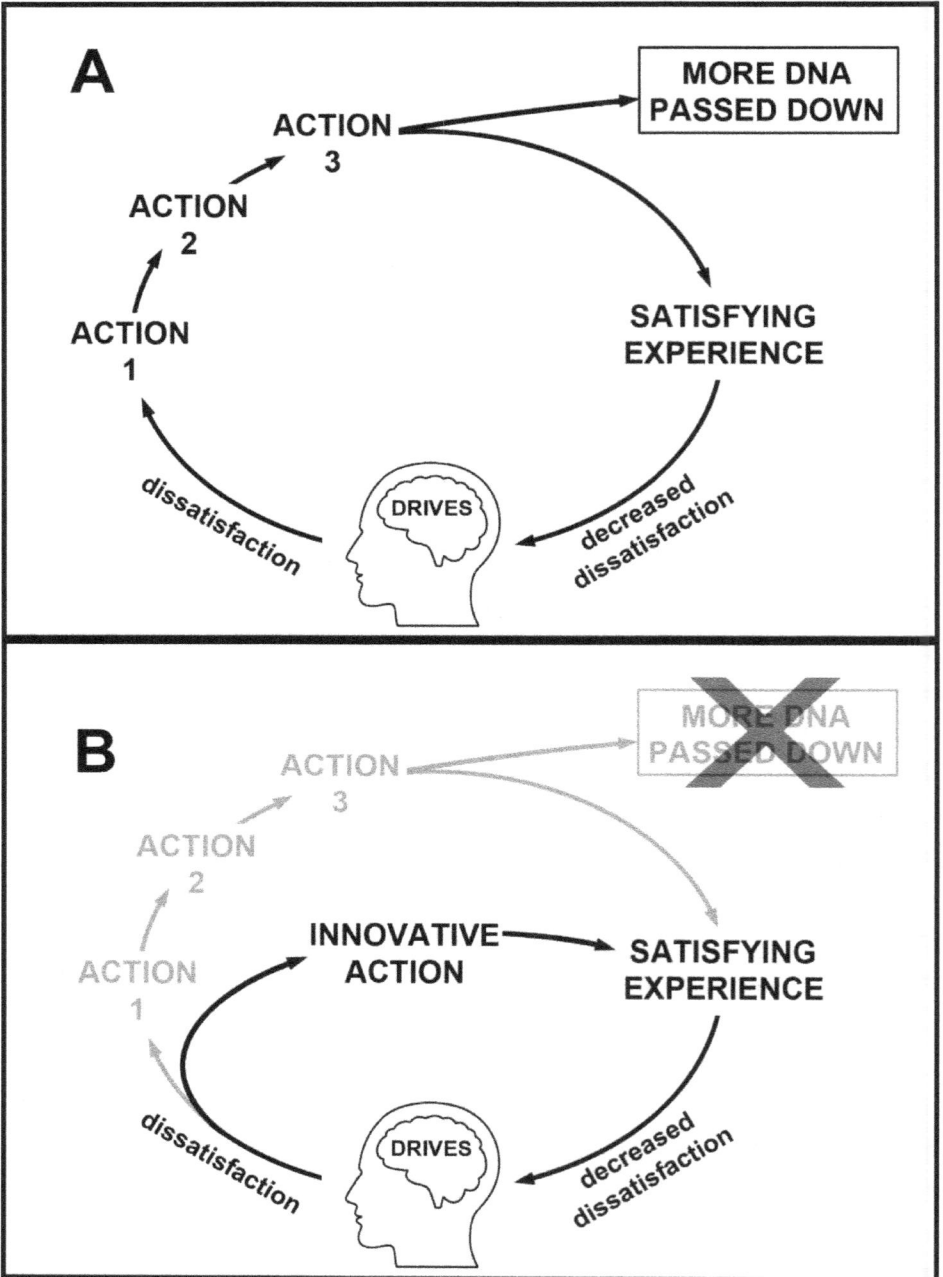

Figure 6: Short circuiting an evolved drive.

These two images show how innovations can lower our evolutionary fitness, simply by offering us easier routes to happiness.

A: In ancient times, our basic neurological drives could only be satisfied if we went to the trouble of carrying out a certain set of actions. As a side effect of those actions, we'd end up passing on more DNA to future generations. Example: A young mother of two finds herself feeling urges to once again have a new baby in her arms, due to drives built into the circuits of her brain. She convinces her husband to agree, and they have a baby. Satisfying that pressing urge has created an evolutionary side effect: her DNA is now much more likely to appear in future generations.

B: In modern times, we can often short circuit these evolved processes, obtaining relief from our brains' dissatisfactions more easily. The new, improved routes to satisfaction often skip steps that might have led us to pass on our DNA, as a side effect. Example: A young mother of two finds herself feeling urges to once again have a new baby in her arms, but she worries that this would be a very impractical decision. She goes onto a contraceptive pill program, then buys a beagle and lavishes care upon it. She loves her pet, and the urge to have a baby slowly fades with time.

innovative behaviors act so rapidly against people that they subject even a privileged modern person to old-fashioned, hard-fisted natural selection processes. Severe drug and alcohol habits offer clear examples of that.

Modern people, and modern nations as a whole, experience the constraints we saw at work on the Second Brewer. Although innovations may offer us thousands of conveniences and luxuries and indulgences, a society won't persist for long if its innovations provide nothing more than that. As with the Second Brewer, the nation and its people have to undergo some strategic suffering; they must invest time and labor in the maintenance and improvement of the infrastructure (roads, farms, industrial plants, etc.) and defense preparedness. Any society that fails to make these continuous investments falls into the unstable condition that we call 'decadence', and soon crumbles under internal and external stresses. In the short run, the necessity of maintaining productivity and military readiness prevents the bulk of society from slipping into a state of pure indulgence, though some citizens certainly do exactly that.

We have found, then, a comforting bulwark that keeps modern societies from slipping willy-nilly into decadence. This is not a very original observation, of course. Most descriptions of economics begin with the assumed axiom that a nation has to maintain its infrastructure and defense capabilities if it hopes to survive for any length of time. Even in a nation where most of the economy focuses on frivolities and luxuries, the system won't persist unless *someone* delivers the mail, fills in potholes, and patrols the borders. However, when we get to Chapter Eight, we're going to see that automation and other improvements are making it ever more possible to have our cake and eat it too.

Before we go there, however, we need to see what effect this short circuiting of drives is having on our evolutionary fitness. Is there a measurable link between the quantity of a culture's embedded innovations, and the tendency of its people to pass on their DNA?

7

LOSING THE URGE

Our direct drives to reproduction can be short circuited, just like the other drives that motivate our behavior. Not only can we find novel ways to eat, to move across the landscape, to entertain ourselves, and so on – we can also adopt innovations that change our habits of reproduction and the way we raise our children. In these latter cases, the link between the innovation and its long-term, evolutionary effects (in terms of natural selection) is particularly obvious. If any species changes its habits in regard to reproduction, we're going to have to *expect* a change in its tendency to pass DNA to future generations.

Let's start with our best known and best studied reproductive drive: the libido, or sex drive. First, let me reassure the reader that there's no indication that we human beings are losing our sex drive. As far as I know, not one serious researcher has ever proposed that our falling worldwide fertility rate, often called the 'fertility crisis', is the result of a declining interest in sex. In fact, it seems more likely that the amount of sexual activity in wealthy countries has increased over the past century or so, right at the same time that voluntary fertility rates have been plummeting. As prosperity has increased, decade by decade, it has given us plenty of new and innovative routes toward sexual gratification. To cite a few familiar examples, there's the extraordinary success of the video and Internet pornography industries, and the popularity of erection-enhancing drugs. More than that, many wealthy societies have actually experienced 'sexual revolutions' – and all of this has happened while birth rates continuously decline.

By its very nature, a sexual revolution consists of a society-wide relaxing of traditional restrictions upon people's access to various forms of sexual gratification. Opportunities for engaging in sexually gratifying acts have been increasing during the past several generations, even as the number of embryonic conceptions has been declining. That suggests that *something* has been short circuiting our most essential drive to reproduction, the libido.

The libido can be short circuited by any act that is sexually gratifying but doesn't increase the probability of pregnancy. The most straightforward innovations for causing this short circuit are contraceptive techniques and technologies. With effective contraception, even the act of delivering viable sperm onto the cervix of a fertile woman on the fertile day of her monthly cycle is reduced to biological irrelevance. Effective, convenient contraception swept the world after the introduction of the birth-control pill in 1961. That's a key historical example of an innovation that decreases human dissatisfaction, while lowering the fertility rate (and thus the evolutionary fitness) of those who use it.

In addition to contraception, there is also a much wider public acknowledgment of the acceptability of sexually gratifying acts that have no possibility of causing pregnancy. Many of these acts were illegal in many parts of the developed world just a few generations ago, and a few such laws still persist. But it has become widely accepted, even among social conservatives, that non-copulatory acts can constitute genuine sexual intercourse. An extraordinary historical event made this issue – which had previously been somewhat taboo in the United States – a topic of household conversation, and one that could be easily grasped by anyone who watched television or read the newspaper. In 1998, U.S. president Bill

Clinton told Congress that he didn't believe that fellatio constituted "sexual relations" between a man and woman. Congress accused him of lying, and impeached him as a result.

Biologically, of course, Clinton was exactly correct. Biological sex is an act of genetic recombination. No act, regardless of its mode of neurological satisfaction, has anything to do with biological sex if it can't lead to the conception of an embryo. But biological sex has almost nothing to do with how we use the term 'sex' in modern wealthy societies, and so the congressional interrogators were also correct, and the public easily under-stood that President Clinton was in fact lying. Clinton's debacle proved to us that fellatio *is* sex, even though it is irrelevant to the biological realities that lie behind sex. We can now assume that the word 'sex' must be defined, at least in American English, as: "acts that gratify the libido." The matter has been settled, effectively, by an act of Congress.

Sigmund Freud was way ahead of everyone else in recognizing (like Clinton's impeachers) that pretty much everything can be regarded as sex if it *feels* like sex... and pretty much anything can feel like sex to somebody, sometime. Freud used the term 'polymorphous perversity' to refer to the malleability of an infant's mental state, in which effectively anything may be identified for use in adult life as a means of sexual gratification.

There's no dictionary word (in any language) that specifically means, "sexual contact between a fertile man and a fertile woman that includes ejaculation inside the vagina without the use of contraceptives." So, for our purposes here, let's invent a term to describe that, and call it BRS: 'biologically relevant sex'. The reason for picking out that very narrow subset of human sexual activity and giving it a name is that it can also be defined in another way, namely as: "the part of the natural human life cycle in which sperm cells are delivered within swimming distance of an egg." Unless we hire laboratory services, BRS is the only event in the world that can lead to the creation of a new child. Non-BRS sexual activity isn't actually a link in the sexual life cycle – not in our species, nor in any other.

It's interesting to note that, since the libido consists of a reward system found in the human brain, we modern people have ended up defining the word 'sex' entirely in terms of neurological activity, rather than in terms that refer to actual reproduction. That doesn't mean we've made any mistakes, however. In legal, psychological, and everyday usage, that definition is correct. BRS is certainly not a sufficient definition of what sex is, except maybe in the context of a biology lab.

The first thing to notice, now that we've got our terms straight, is that BRS only accounts for a tiny fraction of a percent of all sex acts that occur in modern, developed countries. If we compare the number of libido-gratifying acts that occur each day in wealthy nations versus the number of embryos that are conceived, the difference is immense. What that means is that our societies have managed to pry loose the libido (the neurological drive that originally evolved to cause conception of embryos) from the actual act of conception. The libido is a drive that can be satisfied in hundreds of different ways, and it turns out that it can be satisfied *completely* without ever leading to conception. When we do that, we've short circuited our most basic drive toward reproduction. I'm certainly not saying that there's anything wrong with that, but it does lead to some remarkable social consequences, as we'll see.

Another crucial category of neurological drives toward reproduction are the ones that make people adore babies and small children, and want to care for them. I mentioned in Chapter Five that the very sight of a baby causes a chemical reward to be released inside the brains of childless women, and probably everyone else as well. The existence of that sort of reward system inside the brain provides our species with a straightforward, pleasure-based motivation for childcare. That's important, because childcare requires a much bigger investment of labor and resources than the other basic components of reproduction (namely, BRS and pregnancy).

There are several ways of short circuiting the evolved drives toward childcare, and the most direct of these is adoption. Adoption doesn't decrease the amount of investment required to raise a child, but it does separate the act of childcare from the process of passing DNA on to future generations. Adoption is contrary to naive Darwinian expectations, but makes sense because our evolved behaviors are driven by the pleasure centers of the brain. The brain's circuits that cause us to want to care for children originally evolved to increase our chances of passing on our family's DNA. Adoption decreases the dissatisfaction that those circuits create, but disconnects the satisfying feelings of childcare from their original, evolved purpose.

The urge to adopt is very common. In 2008, fully 3% of the children who entered US families did so through adoption rather than birth. This is so, despite the very high monetary cost of adoption, a cost that must be added to the normal costs of raising any child. The US Department of Health and Human Services estimates that a US couple hoping to adopt a child in 2011 could expect to pay up to $40,000 during the process. Furthermore, though no rigorous statistics on waiting times are available, the couple must endure a long wait before their wish is granted – typically six months to two years. The fact that this 'waiting line' is so long suggests that, despite the costs, the demand for adoptable children actually exceeds the supply. All of these numbers surely sound depressing to prospective adopters, but they also indicate a powerful social force: a lot of people in the US and many other wealthy countries want very badly to have a baby in the house. Among these people, this child-rearing desire clearly need not have anything to do with some evolutionary drive to get their own DNA into future generations, because it can be satisfied by the adoption of an unrelated infant.

Since the adoption of unrelated babies has no easy Darwinian explanation, we might assume that the phenomenon ought to be unique to modern humans. We, after all, have the distinction of being the only population of living things in the natural world that lives outside the post-Darwinian threshold. So it's a bit of a surprise to find that non-kin adoption is seen in a number of non-human species, too. One species, the emperor penguin, adopts unrelated chicks quite frequently, and chickless females will actually fight over which of them gets to adopt an orphan. A large 1993 study of an emperor penguin colony found that over 17% of all chicks were adopted at one point or another – a much higher percentage than is seen in human populations. The adoptive parent was almost always a female with no chick of her own, as might be expected. More surprising is the fact that the adopted chicks weren't always orphans. Over half of all adoptions were the result of kidnappings, in which a needy female snatched or lured the chick away from its biological mother.

Obviously, there must be direct, reward-based drives inside the brains of emperor penguins that are satisfied by giving care to the young of their species. We can safely assume that these drives evolved because of the benefit such drives give to each family lineage, by encouraging each female to lavish care upon her chick. Doing so is the only way that she and her mate can pass their DNA on to future generations. But the reward that's released in her brain doesn't come from some abstract knowledge that DNA is being passed along. The reward is more immediate, and comes from the sight, sound and feeling of holding and feeding that fuzzy little chick. Short circuiting a drive like that is actually quite easy. A female might prefer to raise her own chick, but if she doesn't have one, then any chick will do a pretty good job of releasing the chemical rewards in her brain.

In many species, the neurological drive toward caring for offspring can even be short circuited by adopting the young of a different species. There are a lot of documented examples of this phenomenon, which is known as 'cross-species adoption'. The most common examples involve the exploitation of the adoptive parent by the biological parent, and these cases are called 'brood parasitism'. Cowbirds, black-headed ducks and many cuckoo species lay their eggs in the nests of birds of other species, and the host parents adopt the chicks when they hatch. Many factors might be at work in these odd, interspecies relationships, but clearly, one of those factors is a short circuiting of the care-giving drives in the brains of the host parents.

Not all cross-species adoption is parasitic, and the spontaneous, non-parasitic cases are more interesting in the current context. Unfortunately, most of the non-parasitic, 'voluntary' cases of interspecies adoption that are on record have taken place among captive or domesticated animals. Such cases are hard to describe scientifically, because the behavior of captive animals is often quite different from what exists in the wild. In captivity, many animals find themselves living beyond their post-Darwinian threshold, enjoying a constant abundance of food, complete protection from predators, and so on. This can easily cause short circuiting of evolved behavioral drives, just as it does among civilized humans. But even if such cases can't be effectively studied by science, they are also too common to be completely ignored. Many examples have been widely disseminated on the Internet and other media. The most famous of these is the case of Koko, a captive female gorilla that cared for a series of housecats, treating them very much the way we humans treat our pets – though she also attempted to nurse them, as if they were baby gorillas.

In the wild, on the other hand, we rarely see cross-species adoption among animals, but cases do sometimes occur. The most famous example is probably a lioness that adopted a juvenile oryx (a type of antelope) in Samburu Game Reserve, Kenya, in 2002. She cared for the oryx calf for two weeks before it was eaten by another lion. The adoption was widely reported by media around the world, and documented by a film crew.

By now, the reader may have noticed that we're talking about a class of behaviors that's also found in humans. When a person lavishes care upon an individual of another species, giving it food and protection and other services that would be typical of parenthood, then we say that the person is keeping a pet. Pet ownership is thus a form of cross-species adoptive behavior.

Many childless couples lavish care upon household pets, and when they do, they give us a clear example of a short-circuited reproductive drive. Remember: that last phrase doesn't imply that the pets are the cause of the couple's childlessness; it only means that the pets offer satisfaction of a motivational drive, without helping to pass on the couple's DNA. At any rate, childless couples are not the only ones who experience this short circuiting. A mother who lavishes care upon two children and a collie is shunting some of her evolved care-giving behavior away from her DNA lineage, and indeed away from her species. Once again, there's nothing wrong with that. In developed nations, resources are easily sufficient to allow a typical parent to support pets while still providing lavishly for his or her children.

The comparison between loving a pet and loving your own child might seem ludicrous to some readers, but to other readers it will seem both obvious and natural. For those in the former category, note that recent psychological research has shown that the infantile appearance of cute pets does, in fact, elicit the same types of emotional response from people as does the appearance of a human baby. This emotional response is not merely a happy accident. To some degree, it is actually manufactured – that is, it's the result of innovation. Common pet species are often bred selectively to cause them to show infantile qualities as adults, in order to increase the feelings of gratification we get when we offer them care. An adult Pomeranian lapdog, for example, is a full-fledged member of the wolf species, *Canis lupus*, but shows very few of the usual adult characteristics seen in wild wolves. Rather, it has been bred for centuries to grow up into a small, soft-furred, docile adult, maintaining a lot of the physical and behavioral qualities of a young wolf pup throughout its whole life. These infantile qualities greatly improve its popularity as a house pet when compared with the wild-type wolf, which is rarely invited to lie around on our sofas.

It may be interesting, in passing, to point out that even the word that we use in English to describe the babyish qualities of things that aren't babies (such as many pets) is a recent innovation. The word 'cute' is only a few generations old, and initially didn't mean baby-like, but rather meant sexually attractive. Although the word is still used that way, by the mid or late 19th century it had developed its familiar secondary meaning, which describes a positive emotional reaction of the sort that we have toward babies and baby surrogates such as pets, dolls, etc. The fact that there was no word in English that meant 'cute' until a few generations ago may indicate that adults before that time saw little similarity between the emotional reactions evoked by a human baby and the reactions evoked by other things such as household animals.

Another indication that this might be the case is seen in the changing social attitudes in wealthy nations toward animal rights. During the world's first debate on the subject of a proposed animal rights act, England's Martin Act of 1821, the only animals under consideration were horses, cows and sheep. When one of the opposition pointed out, as a *reductio ad absurdum*, that if horses were given protection against cruelty today, dogs and cats might be receiving such protections in the future, the idea was so absurd that the parliamentary chamber exploded into laughter. But in 2004, a California man was given a life sentence for killing a dog. Only eight generations passed between a dog's rights being a joke and being a matter of the gravest legal consequence. Today, at least in the US, it's common for services that seek to place abandoned pets with new owners to call this act of pet placement, "adoption."

All of this suggests that pets serve modern society as an innovative way of short circuiting our neurologically based childcare drives. If that's so, then we'd expect to see people lavishing more time and money on their pets in recent generations, at the same time that they've been having fewer children. The two phenomena should occur in tandem.

Statistics show that that's exactly the case. Pets have become ever more popular during recent generations in wealthy nations, even as fertility rates have fallen. With each passing year, the average American spends 3.1% more on his or her pets (adjusted for inflation), and that's been true for decades. Real spending on pets by individual Americans is doubling every generation.

Pets short circuit our childrearing drive by serving to satisfy our care-giving desires while having nothing to do with biological reproduction. As biological-baby surrogates, they act as short circuits to the childcare drive, much as birth control pills short circuit the libido. Pets and the Pill don't lower our love of children, nor our sex drive, but they offer more efficient, innovative behavioral mechanisms to satisfy those drives. The new, better methods in both cases have this much in common: they make us feel better without leading to the production of children. In both cases, it's clear that the original purpose of the underlying motivational drives was to encourage us toward exactly that goal.

The short circuiting of the libido and childcare drives are the most conspicuous examples of innovation causing a decline in fertility, but they are far from being the only ones. All of the motivational drives built into the circuitry of my brain evolved, originally, to get my family's DNA into future generations. But I live in a world composed of a rich collection of embedded innovations, a synthetic world, and my habits and tendencies and desires are not those of my ancient ancestors. Not by a long shot. I do certain actions over and over (like putting on my clothes every single morning), and others I do once in a lifetime (like visiting the Grand Canyon). It would be possible to list thousands of specific types of action that I have performed in my lifetime... and very few of them would be on a similar list made for one of my ancestors in 10,000 BC. Almost everything on the ancient list was intended to pass on the family's DNA, while hardly anything on my list is relevant to that goal. I live beyond the post-Darwinian threshold, so I feel no pressure. Or, to be more precise, I feel various pressures all the time, but they're almost never directly relevant to biological, evolutionary reality.

In civilized societies, the cause-and-effect relationship between accumulating innovations and decreasing fertility works through the following three steps: 1) Innovations are added to society specifically to decrease dissatisfaction in human brains. 2) These new, easier routes to personal satisfaction short circuit the drives found in the brain – that is, they disconnect them from their evolved purpose, which was to make us have children. 3) The reduced effectiveness of these drives creates a social trend toward voluntary, apathetic infertility. We can call this three-step link between accumulating innovations and falling fertility: **satisfaction sterility.**

We saw, back in Chapter Three, that the accumulation of innovations in a society can be observed as a rise in prosperity. That turned out to be a great boon to us, because it gave us a way of estimating the density and quality of innovations in a society using a single number that's pretty easy to determine: the per capita GDP. Now we get to reap the rewards of that discovery. If satisfaction sterility is caused by accumulating innovations, then we should see more of it in those societies that have higher per capita GDP. This means we can easily observe and actually *measure* satisfaction sterility: it will show up as a correlation between rising per capita GDP and falling fertility rate. Both of those things are being recorded all the time by governments and non-governmental agencies all over the world, so we can get our hands on huge amounts of reliable data, any time we feel like it.

Let's take a closer look at satisfaction sterility, starting with the following observation: Innovation is the only thing that can short circuit human motivational drives. Because that's so, innovation is the sole cause of satisfaction sterility. Remember that we saw, in Chapter One, that human innovation processes don't occur as the result of natural selection tinkering blindly with our DNA. That *is* the way innovations like spider webs first appeared – it's even the way the physical circuitry of our brains was innovated. But human behavioral innovations such as light bulbs, pizzas and refined cocaine are the products of a purposeful ('teleological') process that involves imagination, and planning in advance.

Because that sort of process is *not* constantly subject to natural selection, the only thing that determines the success of the outcome is whether or not it pleases us. If it decreases human dissatisfaction, it's a winner. If an innovation pleases us and is also good for our fitness, as in the case of agriculture, it persists in human society. But if it pleases us and is *bad* for our fitness, as is probably the case for crack cocaine, it nonetheless persists in human society. The things that fail to persist are the things we don't like – regardless of whether they're good or bad for our evolutionary fitness.

Since innovation is the thing that short circuits our drives, we're now in a position to estimate the extent of that short circuiting, based on some very simple and easily observed numbers. Remember, the density and quality of a nation's embedded innovations is strongly indicated by per capita GDP. That means that the average citizen of a nation with a high per capita GDP should be experiencing a lot of short circuiting of his or her evolved neurological drives. And, as we've seen, that short circuiting process should lead to a single, easily measured outcome: it should lower the tendency of the average citizen to voluntarily produce children. A large accumulation of innovations means that a society is offering lots of pleasant opportunities to short circuit the urge to have children, and that means that we should see lower fertility among nations with higher GDP.

Of course, that observation only applies to modern, wealthy nations. We saw in Chapter Six that when a society exceeds a certain level of prosperity (the post-Darwinian threshold), it loses the link between resource availability and reproductive success – a link that characterizes almost all living species. Below that threshold level of prosperity, the citizens of human societies (civilized or not) are subject to natural selection.

But by the argument we've been following, we can now say that the opposite trend should exist in rich societies. Among the nations that are wealthier than the post-Darwinian threshold (about $5000 per capita GDP), rising prosperity should short circuit the drive to have children, causing a decline in fertility.

I suppose I should admit that this 'prediction' is kind of an unfair set-up, because the reader probably knows that the statement is true in advance, even without seeing any actual data. One of the most remarkable aspects of modern civilization is that people who are relatively lacking in socioeconomic power can in fact have a great number of children if they wish to. In some civilized societies (such as those of Europe and North America), the poor produce more children per capita than the rich – a flagrant violation of Darwinian principles. For example, according to data from the U.S. Census Bureau, a woman born in 1958 who lived in a household with an income of $20,000 a year (standardized to 2014 dollars) had an average of 2.5 children during her reproductive years, which were at their peak during the 1980s. But if she was living in a house that made $80,000 a year, she had only 1.7 children. Statistically speaking, the more children an American couple can afford, the fewer they have.

There are plenty more numbers like that, to back up the assertion that rising per capita GDP is linked to falling fertility rates. The trend is global, long-term, and extremely well documented. If you'd like to see some data and sources, have a look at Appendix One, especially at Figure A2 and the text that goes with it.

Consider two households: the first consists of an impoverished couple living in a tenement apartment with seven children, and the second consists of a billionaire couple with one child. The Darwinian explanation (which doesn't work) would be this: the rich couple is producing fewer offspring because they are investing more resources to ensure that their child is fully protected and empowered, thus maximizing his or her chance to survive to adulthood and reproduce. That argument, which works with amazing accuracy across hundreds of other well-documented species, simply doesn't apply here. True, the single child of the billionaire couple has a nearly 100% chance of surviving to reproductive age in good health. But we can't say that we really expect *fewer than one* of the tenement couple's seven kids to grow up and have kids of his or her own. We can reasonably expect that, out of seven, at least four or five are going to grow up and have the opportunity to reproduce; in fact, it's fairly likely that all seven will do so. That means that the poor couple has *much* higher evolutionary fitness than the billionaires. The mathematical logic of population genetics leads us to that conclusion with no possibility of escape.

The correlation between increasing per capita GDP and decreasing fertility in civilized societies defies simple Darwinian reasoning, but makes perfect sense when understood in terms of embedded innovations and short-circuited drives. Rising prosperity means innovative new methods of obtaining personal satisfaction, and these often short circuit the evolved drives that pressure us to do the behaviors that pass on our DNA to future generations. This link between rising per capita GDP and falling fertility is satisfaction sterility.

Some readers may be baffled by all this talk of infertility, because it's a well-known fact that the world population is still rising – and rising fast. The global population grew bigger by 77 million people in 2014, according to the US Census Bureau. That's a huge number; 77 million is roughly ten times the number people who ever existed on earth at any one time, during the first 90% of human history. By any reasonable, biologically relevant standard, our global population is exploding out of bounds. Why, then, should we be concerned about 'satisfaction sterility', or any other supposed trend toward infertility and low birth rates?

The numerical increase in global population is a statistical illusion. Our exploding population was a real issue for a very long time, and that was true even just a generation or two ago. Today, yes, population is rising, but not due to any factors that will prove important over the long run. The raw number of people alive on earth is only growing from day to day because the elderly demographic sector in each nation is filling up. When that filling is done, death rates will catch up with the steadily falling birth rates seen around the world. Human longevity has increased, true, but all of us who are currently living will *eventually* die. The issue of 'exploding population' was the most important ecological and social-biological issue from the time of Malthus in the 1790s until roughly the 1970s. Today, the explosion is over, and the only reason the numbers are still expanding is because the dust of the explosion hasn't settled yet. The collapse is about to begin.

Here are the numbers, in a nutshell: As of 2015, the world population is rising at a rate of just over 1% per year. If we held to that rate steadily, our population would *double* by the year 2080! Any alarmist who's worth his salt will tell you that frightening statistic... but every serious population scientist will tell you not to worry about it. The reality of the situation is that our rate of population growth is declining, creating a force of deceleration that's similar to stepping on the brake pedal of a car. The key factor to consider in this regard is the number of children that the average woman produces in her lifetime, a factor known as the 'total fertility rate'. According to the United Nations, in 1960 the world's total fertility rate was about 4.5 to 5.0 children per woman. Today, that number is estimated at 2.3 to 2.5, and still plummeting downward. When it hits 2.1, our population collapse will officially begin. Our situation is like that of people sitting in a car that has been speeding out of control up a steep hill, and we've just stepped on the clutch. We haven't come to a stop yet... but we will. Then we're going to start rolling backward, down the hill.

The days are past when the big topic in modern population studies (and economic forecasting) was the population explosion. The top matters of discussion now are the fertility crisis and the aging of populations. World-wide, the category of people aged 60 or over was 9.2% in 1990. That rose to 11.7% by 2013, and the UN estimates it will grow to 21.1% by 2050. Where are all these old people coming from? The answer is simple: our ever-improving treasure trove of embedded innovations keeps making our lives better, and one of the many benefits is that we all get to live longer. Almost every nation on earth has seen an increase in life expectancy during the past generation or two. Longer life means a bigger slice of the population who are elderly.

It may not be instantly obvious why an improvement in longevity should create the illusion of an exploding population, when birth rates are actually falling... and falling faster every year. Let me explain how that works, using a simple analogy.

Imagine that you have created a pond by building a dam in front of a small stream (Figure 7, panel A). You've let the space behind the dam fill up with water, and the result is a pleasant, stable pool. Water is constantly feeding in due to a gradual flow from a stream uphill. Each molecule of that water spends a fair amount of time lazily circulating around in the pond, and eventually it spills out over the dam. In this metaphor, the pond represents our global human population and we are the water molecules; the stream that feeds the pond is births, and the spillover represents deaths. For simplicity, let's say the pond starts out in a state of equilibrium, meaning that births and deaths are equal.

Now, over a short period of hard work, you raise the height of the dam substantially (Figure 7, panel B). The raising of the dam is similar to what happens when conditions are improved in a society, so that life expectancy increases. Each molecule of water that enters the pond will circulate for a longer period before it spills out. After the raising of the dam, some time is going to pass before the little stream is able to fill the new, bigger volume of the pond. During that time, there is *no* spillover, and only small amounts of water are lost by seepage or evaporation; the output isn't nearly as large as the input. But eventually, the pond will fill up to the top of the dam again and begin to spill over, and the situation will enter a new steady state (Figure 7, panel C).

Similarly, in a society where life expectancy has suddenly increased, the death rate can be near zero for a decade or more. But death rates and birth rates ultimately have to even out, because everybody who is born eventually dies. The long interim period, the period of filling up, is what's happening in a lot of nations right now. Life expectancies have increased and the first generation living under the new, improved conditions is still alive, gradually getting older. The pond is still filling up. The stream of births is trickling in, but almost no one is dying... yet.

Nowhere do we see this phenomenon more clearly than in the oil-rich countries of the Middle East. The CIA estimates that, in the nation of Qatar in 2013, only one person died out of every 649! Unless Qatar has achieved an average life expectancy of 649 years, that's not a sustainable situation. The condition of the oil-rich nations is a temporary one, like the

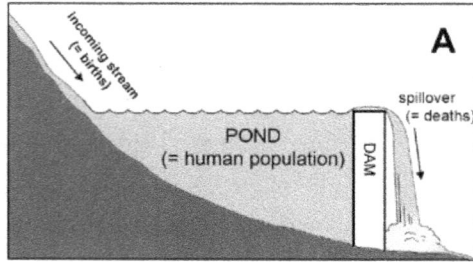

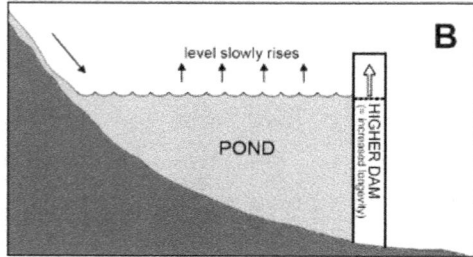

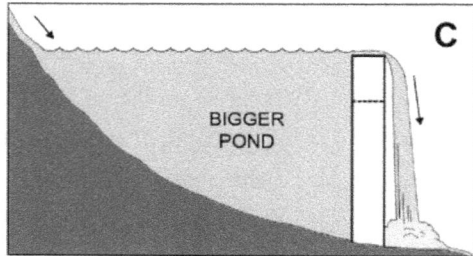

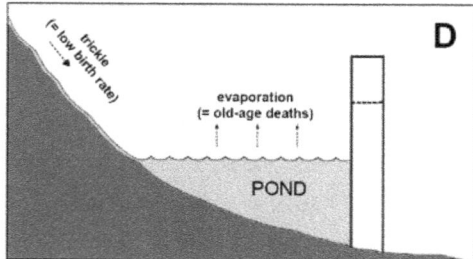

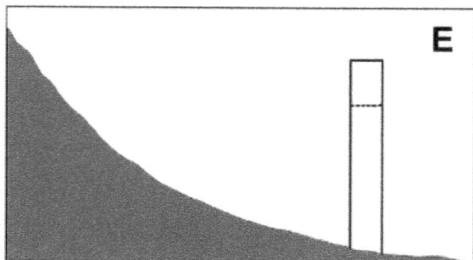

Figure 7. Population can rise, even as birth rates fall.

Here's an analogy, to show how that can be.

A: The global population can be compared to a pond behind a dam. The pond receives water from a stream (representing births) and loses water over a spillway (deaths). If the input and output are equal, the pond's volume stays the same (stable population size).

B: If the dam's height is quickly raised, a long period may follow in which there is no spillover, because the pond hasn't finished filling yet. The raising of the dam represents increasing people's life expectancies. That is to say, a molecule of water entering the pond can now expect to stay there a lot longer before it spills over the dam. This situation shows us what's happening in the global human population today: death rates are temporarily much lower than birth rates, even though birth rates are actually falling.

C: Eventually the water reaches the new top of the dam, and the rate of spillover (death rate) becomes equal to the stream's input (birth rate). In terms of the human population, that means that our life expectancies will remain very high, but in a few decades our death rate will nonetheless catch up with our birth rate.

D: (Panels D and E anticipate topics from the remaining chapters of the book.) The stream is quickly drying up now. This represents the collapse of the birth rate, though longevity and quality of life remain high. The death rate just barely keeps up with the dwindling birth rate, and is shown here as evaporation from the shrinking pond.

E: The situation has reached the only possible stable conclusion, and the pond is gone. Our species has gone extinct, like millions of other species before.

period right after a dam has been raised to a higher level, while the pond behind it hasn't yet filled to the top. In the case of human populations, the reservoir that's still filling is a new, large demographic subpopulation that consists of the elderly. Qatar is an extreme case, but it is not alone: similar situations are occurring around the world. The result is that global population is rising (for the moment) even though birth rates are falling quickly. So the world's rising population is little more than a numerical illusion; it has almost no bearing on our future.

There's another thing to notice about the pond metaphor before we leave it. The water level will rise behind the new, improved dam even if something happens upstream that begins to slowly choke off the supply of water to the pond – in other words, something that begins cutting down the rate of births. If that happens, we can step back and look at the whole situation (pond, dam, stream) and understand that the pond is eventually going to dry up. And yet, at the moment, we don't see anything of the kind. All that our eyes reveal to us is that the pond is growing larger every day, behind its new dam. The only way we can *see* that the pond is on its way toward drying up completely is to use our powers of understanding to take in the big picture.

The big picture is this: It doesn't matter how well-built the dam is, or how placid and long-lasting the water, or even that the pond is growing measurably larger every day. In the long run, only one thing is going to matter, and that's the rate of input. The inflowing stream is slowly drying up, and when it's gone, the pond is going to disappear completely.

We've established that rising prosperity is linked with falling birth rates, and that this is not merely coincidental. They are tied together as cause and effect. This causal linkage is not necessarily any big deal, in itself. For one thing, any number of natural processes might serve to regulate it and keep it in check. Even if that's not the case, we humans are far from helpless. We can reverse that trend any time we please, and we have quite a number of different means for doing so at our disposal.

So as we come to the end of Part II, we haven't yet really defined any crisis that's looming in our future. We've merely put a finger on an interesting social issue, one that has connections to economics, evolution and behavioral neurology. Still, if I find a small lump under my skin today, even if science can tell me exactly what it is, nonetheless the thing *may* turn out to be unstoppable. Just because a problem starts small, that doesn't necessarily mean we can do anything to stop it – and that can be true even if we can see all of its conditions and parameters in advance. The rest of this book will be devoted to making a prognosis. What threats does satisfaction sterility present to our species? What (if anything) can we do to keep it under control?

But before we get into all of that, we'd better have one more look, just to make sure that satisfaction sterility is really the cause of the problem. Could something else be causing birth rates to fall? What if the main cause isn't actually the easing of dissatisfaction that we see in wealthy nations, but some other, coincidental factor? Let's briefly consider the possibilities, if only to get them out of the way.

The first obvious thing to check is the phenomenon of *involuntary* infertility. The whole argument of this book so far has been aimed at explaining why people are having fewer children, and that carries the hidden assumption that we humans are doing it on purpose, at least on an unconscious level, due to an odd sort of apathy. What if the problem is really a medical one? A number of alarming news reports have popped up in recent decades, suggesting that falling sperm counts are a growing problem... maybe even a global one. That's an eye-catching and sensational claim, and if it's true then it might go a long way toward explaining the falling birth rate. After all, industrial societies don't just accumulate wealth and innovations; they also accumulate chemical pollution, urban stress, and other environmental factors that might disrupt our sensitive reproductive systems.

Here's what we really know about the matter. First, biological infertility is certainly a real phenomenon. Best estimates suggest that roughly 3% to 7% of couples are biologically infertile, worldwide. But those estimates are rough indeed, and that means it's not really possible to say whether the global biological infertility rate is stable, increasing, or decreasing. Still, that doesn't mean that all those media reports are just a bunch of baseless rumors. Each news story usually echoes the findings of some particular, newly published scientific study on male fertility in one country or another. If one study finds no sign of change, then it's not news, so it doesn't end up in the papers. But if another study shows falling sperm counts in some region of (say) Europe or the US or Japan, then *that's* news. So if we want to find out the broader truth of the matter, what's needed is a comparative review of all such studies, undertaken by a reviewer who is also a reputable scientist. Fortunately, just such a review was published in 2008. We can now say with confidence that there is no evidence of any global decrease in sperm counts, despite all the news reports to the contrary.

Even if we hadn't been able to get our hands on a sober assessment of that situation, we could still have been pretty sure that involuntary infertility is not a sufficient explanation for the world's falling birth rates. The decrease in birth rate among wealthy nations is too rapid. There's no way such a steep decline in average family size could be accounted for entirely by a rise in physical fertility problems, unless the latter were caused by a plague of previously unknown proportions. In that case, trust me, we'd all be well aware of it. The problem isn't a lack of viable sperm cells and working ovaries, nor is it a lack of sex drive. We modern people simply want fewer children. The richer our nations become, the stronger our silent resolve on that issue becomes. That phenomenon – the link between rising prosperity and falling birth rate – is satisfaction sterility.

Satisfaction sterility is not a behavioral malaise, it's a motivational one. The problem isn't that people are failing to act on their desire to have children; the problem is that the desire *itself* is decreasing. A person like you or me, who lives inside an innovation-rich, civilized environment, experiences evolutionarily novel ways of satisfying his or her motivational neurological circuits. This tends to make us lose interest in those activities which the brain systems originally evolved to reinforce.

Nonetheless, few people who exercise voluntary infertility *believe* that what they're doing is succumbing to satisfaction sterility. Many people, even healthy young adults in the richest societies, honestly believe that they aren't in a socioeconomic position to have a child. But there can be a huge gap between our feelings and what we can rationally deduce to be true about ourselves, when we step back and consider our situation with a bit of perspective.

Survey interviewers in Europe, the United States and other Western regions often ask people what they regard as the optimum number of children in a family, but they rarely ask the follow-up question, "Why not have more?" That's too bad, because we could sure use data on that topic, including self-reported, subjective data. But fortunately, in many other parts of the world, including a lot of East Asia, that question seems like a perfectly natural thing to ask somebody. Many newspaper surveys in Japan since the mid-1990s have asked exactly that question of thousands of young people, and have reported that most of them feel that the main impediment to having more children is the expense. For example, one large survey in Japan in 2003 reported that 75% of respondents felt that way. In Chapter Eight, we'll see that China has now also been the site of such surveys, and a majority of recent respondents have said the same thing.

Consider the absurdity of that. Imagine going around to interview the young generation in some nation. You know that they're having fewer children than any of their ancestors, so you ask them: "Why?" You also know in advance that you're speaking with the richest generation in all of that nation's long, long history... and yet, they tell you that their reason for not reproducing is: "I can't afford to have children."

As absurd as that may be, I think everyone has a gut understanding of the apparent sense of the statement by a modern young person: "I can't afford to have children." We all know that prevailing social mores require that a responsible parent be able to afford a long list of amenities and services to support a child. More than that, having a child when you can't afford to fulfill all those obligations is almost universally considered grossly irresponsible. The list of expensive "requirements" consists of a dense collection of embedded innovations that have arisen in civilized societies in recent generations. The reader can see that this is so by simply writing out a list of what you're going to have to spend money on, when you have your next child. Then read the list, and ask yourself how many of those things were available to our ancestors in 10,000 BC... or even just 200 years ago.

These innovations have been established to improve our children's access to the entitlements of civilization (education, etc.), to improve their security (medical care, etc.), to improve their comfort (consumer electronics, etc.), and to improve the convenience to parents of raising a child (daycare, etc.). Still, however grudgingly, we must admit that in the strictest sense, none of these innovations can quite be regarded as necessities, because in 10,000 BC, before any of these things existed in modern form, people did manage to have children. That entire kit of innovations has not only appeared in our societies since then, it has gone from being desirable to being perceived as absolutely necessary... so much so that a substantial proportion of a national population can state frankly that they would rather forego their natural inclination to have children than produce children without access to the entire kit. That's what a young, middle-class person really means by, "I can't afford to have children."

This might be a good moment to remember the 'drug tolerance' analogy for the satisfactions afforded by innovations, as discussed in Chapter Four. We humans have a nearly infinite capacity for getting used to new, improved conditions, and learning to take them for granted. We easily make the shift from feeling that we're satisfying a *desire* to feeling that we're fulfilling a *need*.

As far as childrearing is concerned, our ancestors – even those just ten generations ago – would shake their heads in dismay if told about our modern attitude. They would surely scoff at our claim that "it's expensive to raise a child," and might be inclined to emphasize a different aspect of the situation, namely that the *drive* to have children must be pretty weak in our modern generation. A lot of us today, when confronted with the discouraging prospect of dealing with all the expenses and difficulties of having a new child, find ourselves perfectly capable of just skipping the whole thing. To most of our ancient ancestors, that decision would have been inconceivable. In fact, that's the only reason you and I are here.

Part Three

OUR PROSPECTS

8

NATURAL PROTECTION

Parts I and II showed us that civilization provides us with a continuous stream of innovations that directly or indirectly stimulate our brain's pleasure centers, short circuiting the drives that make us want to have children. We've seen that fertility rates are falling, worldwide. We've seen that we can measure the link between our ever improving stock of innovations and our declining birth rates quite easily, because any culture that is becoming more saturated with embedded innovations has a rising per capita GDP. In fact, to say that a society's accumulation of embedded innovations is improving (in density and quality) is really just another way of saying that economic development is occurring.

Still, all we've really done so far in this book is explain what causes the observed link between rising prosperity and falling fertility, and to give that link a name: satisfaction sterility. The link is found in the physical circuitry of our brains, which evolved to give us one set of behaviors, but is now being applied to very different uses. These new behaviors decrease our dissatisfaction more efficiently than the old ones, 'short circuiting' the underlying drives.

In itself, none of that is cause for alarm. Just because I tell you that there's an open flame in my house, that doesn't necessarily mean you should call the fire department. Maybe I'm just describing a candle on my mantelpiece. But if someone asks me, "Do you mean a small flame, or a big one?" then that person is asking the wrong question. The important question is this: "Is the flame under control, or out of control?" Even a tiny flame can consume a whole house, if it's not kept under control. For the moment, satisfaction sterility is a relatively small part of the human condition... certainly not the equivalent of a raging house fire. But if we want to know whether we should be concerned, then it's not really comforting to know that the matter is still a small one. What we need to know is whether it can or cannot be controlled in the future. If not, then we should expect that today's small problem is going to grow much, much larger as time goes by.

There are actually two separate questions we need to ask, regarding the matter of control, and we'll spend this chapter and the next one dealing with them, one at a time. The first question is: Will the whole thing take care of itself somehow, or does it really have the potential to grow worse and worse if we simply ignore it? Second, if it's *not* going to come to a natural halt on its own, then do we have the power to intervene?

The current chapter is devoted to the first question: Is there a natural limit to human satisfaction sterility? A number of natural phenomena might (in theory) act to slow down the progress of satisfaction sterility, keeping it in check. We'll attempt to investigate every possibility along those lines. As you've probably already guessed, that's not going to work out, and we're going to be left at the end of this chapter with the understanding that satisfaction sterility may be a small thing right now, but it's poised to spread out of all control. In Chapter Nine, we'll turn to the question: What are we going to do about it?

But let's not get ahead of ourselves.

*　　*　　*

The idea that nature has *nothing* in store to protect us from our own satisfaction sterility might seem unlikely, at first glance. Nature is full of checks and balances: elegant mechanisms that regulate everything from our internal organs to the world's ecosystems, keeping everything working as smoothly as clockwork. Let's see if we can't find a natural limit that can regulate the effects of satisfaction sterility, keeping it within reasonable bounds.

The first place to look, surely, is at that great organizer of the living world: natural selection. Natural selection has guided us and our various ancestral species through any number of crises during the past 3400 million years, and if satisfaction sterility really counts as a 'crisis' then it's a very recent one. Remember, it began less than fourteen generations ago. On the other hand, those comforting remarks carry a certain, haunting ring of being "famous last words." Over 99% of all species that have ever existed *have* gone extinct. If they could have talked, then the members of most of those lost species might have said the same thing, as their end was approaching: "Our problem is only a recent issue, and we've been around a long time." Each and every one of those now-extinct species had a biological lineage going back all the way to the dawn of life, just as we do... a winning streak over three billion years long. But in each case, extinction was just around the corner.

Still, it's not totally unreasonable for us to hope that natural selection has prepared something to protect us from satisfaction sterility. We are accustomed to taking it for granted that natural selection has worked out all the bugs in our survival systems. For example, after I've eaten a piece of bread, I take it for granted that my body will have working systems for digesting it. I eat the bread, forget all about it, and let nature do the rest. My lineage of organisms has been here a long time, and all those processes of digestive chemistry were worked out ages ago, by trial and error acting upon an endless series of generations among my ancient ancestors. As a result, I get the luxury of being able to eat a piece of bread without worrying about how I'm going to extract glucose and other useful chemicals from it. Nature's got all that stuff covered. So why shouldn't I expect that the process of natural selection among my ancient ancestors also worked out some sort of system for protecting humanity against satisfaction sterility?

The reason we can't hope for any such thing is precisely that the problem is so very new. Satisfaction sterility was effectively non-existent just fourteen generations ago. Remember, until a society achieves a level of prosperity that puts it beyond the post-Darwinian threshold, there's no such thing as satisfaction sterility. Below a per capita GDP of about $5000 a year (as expressed in 2015 US dollars), most innovations not only decrease the average person's dissatisfaction, they also improve the chances of successfully raising children. There's virtually no short circuiting of the evolved drives in human brains. In that sort of developing society, the set of things that most people desire is almost identical to the set of things that increases their evolutionary fitness. People want enough food to avoid starvation, and a source of water that doesn't give their children typhoid, and so on.

But today, as we saw in Chapter Six, all but the poorest of modern nations are at a level of prosperity that puts the average person beyond the post-Darwinian threshold. In other words, our accumulation of innovations and wealth has grown sufficiently dense that the typical citizen, worldwide, can have and raise as many children as he or she wants. The main limit on family size is no longer hunger, childhood disease, or any of the other limitations that dominated the lives of our ancestors. Since my evolutionary 'fitness' is just a technical term for the number of children I father and raise to adulthood, I am empowered to be as fit or unfit as I choose. Natural selection has no real say in the matter... it can barely even touch me.

The flip side of that delightful freedom is that natural selection can't mend the errors of my ways. If I don't feel like having children, or if I feel like having one child instead of twelve, then natural selection would surely judge me very harshly – if it could get a grip on me. After a few generations, it would weed out my family and replace us with people who were more avid participants in life's basic competition for reproductive success. But if social and economic success cause *everyone* to eventually lose interest in that reproductive competition, then that process of natural selection doesn't actually change anything. My lineage will die off, only to be replaced by other family lineages that also eventually die off for the same reason. In principle, that process can continue indefinitely, until every family has disappeared.

The straightforward mathematics of population genetics shows us that it's not the meek nor the bold who inherit the earth. In each generation, the earth is inherited by the fecund. If you leave behind ten fully grown children, and your neighbors leave none, then the world is yours. But now that we're beyond the post-Darwinian threshold, you'd have to be pretty short-sighted to think that such a victory is going to last. Even if you have a hundred grandchildren, you can't count on your DNA lasting more than a few generations. As your society continues to get richer, finding newer and better short-circuiting innovations, then all one hundred of your grandkids are going to experience some degree of satisfaction sterility. The fact that you were such an enthusiastic reproducer does not ensure that your great-grandchildren will inherit that enthusiasm.

There's a big assumption in that last statement, namely that the personal desire to have a lot of children is something we humans acquire while we're growing up, and while we're living our lives. In other words, we're assuming that a person's choice of family size is *not* an inherited compulsion, passed down as a piece of DNA code from his or her parents, hardwired into the structure of the brain. We're assuming that the human behavioral tendency toward fecundity is acquired, rather than genetic. Is that a safe assumption? Most human behavior scientists would say it is, but still, we'd better have a closer look.

Let's imagine that there really is a gene, a piece of DNA code, that can be passed along in human families, causing people to feel compelled to raise a lot of kids. If you inherit this gene, you just can't help yourself, you've got to produce a large number of children. There's no actual, genetic evidence of any such gene, but that doesn't mean there's not one out there somewhere. Let's call it the 'fecundity gene'. Now let's give free reign to our imaginations, and picture how this supposed fecundity gene might persist and spread in our developed nations.

A dozen generations ago, when almost everyone had huge families (if they could afford them), you would never even have noticed that the fecundity gene existed. Was it common, or rare? There was no way to tell, back then. Families that carried the fecundity gene had a lot of kids, but then, so did families that didn't carry it. Then, gradually, customs began to change, and average family size got smaller and smaller with each generation. Particularly, all of the families got smaller *except* those families (if any) who were carrying the fecundity gene.

So, to find the gene in the modern human population, all we have to do is eliminate everyone who obviously doesn't have it. We ignore the populations of less developed nations, where satisfaction sterility has not yet become evident. For the same reason, we ignore those families in developed nations who have unusually low income level, by local standards. We ignore families that have immigrated to developed nations during the past two generations, because their traditions may not have eroded away yet. We ignore families with religious doctrines that discourage family planning, because their large family sizes may be the result of social pressures. We are left with a sample of hundreds of millions of people (worldwide) who are prosperous, and who have both a family history and cultural history of prosperity. These are the people who are most likely to show satisfaction sterility. Among that sample, average family size really is mighty small... far below the replacement rate. But a small proportion of those families *do* have a lot of children. If anyone is carrying a fecundity gene in this world, that's the place we're going to find it.

We've just discovered two things about the fecundity gene, without even doing any lab work. The first is that it must be awfully rare. If it were common, then during our recent generations, while the fashion for large families has been slowly evaporating, there would have been a big, obvious chunk of the population that refused to follow the fashion trend. Despite sharing the same prosperity level, cultural heritage and religious views as the shrinking families around them, the women of these families would have continued to compulsively produce ten or fifteen children apiece – and they would still be doing so today. If that were a common phenomenon, then by now everyone in the developed world would be well aware of it. But we're not, and that means it must be extraordinarily rare, if it exists at all.

The second thing we can say about the imaginary fecundity gene is this: If it existed, it would have to be spreading like wildfire today. The nations of the developed world almost all have shrinking populations, or are growing only due to immigration. Since local populations are not reproducing fast enough to maintain population size, any supposed gene that was causing certain families to compulsively produce great flocks of children would spread quickly. Here's a concrete example: Picture a city of a million people who produce 1.5 children per couple. There's a fecundity gene, but it's rare – only 1% of couples have even one copy of it in either parent. The gene compels anyone who carries it to have ten children (on average) with his or her mate. A gene like that would spread amazingly fast. Within four generations, over half the couples in the city would carry the fecundity gene. Three generations later, the city's population would hit twenty million, and still be exploding out of all control. If you want to see the actual numbers for this hypothetical case, have a look at Appendix Two, Topic C.

What does that tell us? It tells us that if the fecundity gene is not currently common (and we've seen that it can't be), then it almost certainly doesn't exist at all. This is a huge relief, because such a gene would spread so quickly in the modern world that it would drive the human species extinct much more rapidly than satisfaction sterility, and under much uglier circumstances. In a century or two, we would practically drown under a sea of people.

So, there's no fecundity gene to *force* humans to reproduce. As we've already seen, the main physical circuits in the brain that really lead to fecundity are the libido and the drive to childcare. Neither of those is actually a drive toward fecundity. It's possible – even easy – to satisfy either of those drives without actually making any children. Through almost all of human history, satisfying the libido and the drive to childcare caused fecundity to occur as a side effect. Not anymore. These drives, like so many others, have proven to be easily short circuited by cultural and technological innovations.

But most of us don't hope our species is going to get assistance from anything as abstruse as 'natural selection forces'. Even for those without religious inclinations, there's a general sense that some sort of instinctive drive exists toward the good of the species. We like to believe that we don't *want* to see our species go extinct, and that we are more than willing to act, as individuals, for the good of our species. This belief may even seem like common sense, at first glance. Probably, a lot of people feel that everyone has *some* interest in the good of our species as a result of simple biology, and that there's no need to seek supernatural justifications for those feelings.

Unfortunately, biology doesn't back us up on this one. There's no evidence of any inclination toward the 'good of the species' in any form of life, and there are strong theoretical reasons to doubt that such a thing could exist. Evolution favors drives that tend to get each individual's DNA passed along into future generations. The main impediment to that set of drives is almost always competition over limited resources. Each species is ecologically defined by what it needs, so competition over resources is usually the most powerful environmental limitation experienced by a living thing. And who are the most serious competitors? Other members of the *same species* – because they need exactly the same resources.

Humans are just as subject to these principles as any other species of living thing. Sparrows squabble over millet seeds, and we humans squabble over oil reserves. It is perfectly plausible that some individual persons are immune to these competitive principles, and feel genuine humanism. They feel just as bad knowing that harm is coming to their most ruthless enemy as they would feel if they heard that the same harm was happening to their own children. There's no arguing that humanism of that sort is a genuine love of the species... but it's not a common sentiment.

The most straightforward objection to claiming that humans have a drive toward the good of the species is warfare, and the ethnic and ideological tensions that typically underlie warfare. We are all of us members of the species that carried out World Wars One and Two, among other memorable examples. That's powerful evidence suggesting that the great majority of us only want to work for the good of *certain* people, not *all* people.

Consider the following argument: The vast majority of people on earth categorically hate at least one nationality, ethnicity, race or religion. Imagine telling such a person (and being believed): "In twenty years, the ethnic group you hate the most will conquer the world and kill everyone but their own kind. Let me offer you a button that will activate a time bomb that will go off in fifty years, killing them all in a glorious act of final vengeance." It's true that not *everyone* would push the button to cause human extinction, but a substantial proportion of people would do so. Their love is not for their species, but for some subset of its population... usually a subset that includes themselves.

So, we've established that pure humanism isn't likely to stop the progress of satisfaction sterility. Let's try a different tack, then. Certain fields of science deal directly with population issues. Perhaps one of them will offer useful insights. We can start with a likely candidate: population ecology.

The regulation of population growth is generally a topic that falls in the territory covered by the science of ecology. Over the past century, ecologists have uncovered a lot of regulatory mechanisms that keep populations of one species or another from exploding out of control, or dwindling to extinction. Of course, sometimes populations of various species *do* explode, or go extinct, but there's also a lot of order and stasis in nature. Could natural, regulatory mechanisms work to keep our population at a steady level?

As we begin considering this possibility, it's right for us to feel some blanket skepticism. After all, we humans are a species that has gone from a population of one billion to seven billion in nine generations! What kind of "regulatory mechanism" are we talking about here? One that doesn't work very well, apparently. Still, the idea shouldn't be dismissed until we've given it some sober investigation.

In general, populations of any species are regulated by limitations upon the opportunities for survival and reproduction. As the individuals become crowded they starve, kill each other, glut their predators, transmit epidemic diseases, etc. In natural populations that have been released from most of their predator pressure (like we have) and where individual territorialism is relaxed enough that crowded conditions are tolerated (as with our species), the regulator of population size is usually starvation.

Even starvation won't stabilize a population, except as a result of a long period of natural selection to balance the rate of reproduction against the food supply. Most species have undergone long evolutionary periods that effectively match their reproductive habits and physiology to their food sources and other environmental circumstances. Other species never come to a stable population size, but persist in a constant cycle of population explosions, crashes, and slow recoveries. Locusts and lemmings are two familiar examples. In these cases, the population is limited by periodic starvation, so in a certain sense hunger serves as a regulator, but it never provides the stability that's seen in the majority of species. The populations grow to ridiculous density, eat themselves out of house and home, die in droves, and then the few survivors gradually repopulate the area. A few years later, the whole cycle repeats.

But these cyclic populations are very much the exception. For most species, including ours, a population explosion is not a common, regular event – it's the result of a severe loss of ecological balance. Since the threat of starvation is usually the key to maintaining a well regulated population, a one-time population explosion of such a species is usually caused by an unnaturally rich source of food. When an unlimited food supply appears out of nowhere, a once-stable population of almost any animal species may grow to amazing density in a very small number of generations. Then comes the crash, known as a die-off.

A classic example of this phenomenon was seen in a herd of caribou in the early 20th century on St. Paul Island, one of the small Pribilof islands off the Alaskan shore. The US government introduced 25 caribou to this uninhabited island in 1911, hoping the herd could be managed and harvested as a source of extra meat for people living on the nearby mainland. Alaska Fish and Wildlife Service oversaw the annual hunt, recording the number of animals killed, and estimating how many were left behind. Both numbers rose progressively, until by 1938 there were over 2000 animals living on the island – a ridiculous number for such a small island. During the next few years, the population experienced a spectacular die-off, and by 1946, the annual hunt was discontinued. Still, the population continued to crash, and by 1950 only eight caribou remained.

A mammalogist, Victor Scheffer, did an ecological study of the St. Paul caribou herd. He reported that the rate of replacement for the main foodstuffs (grasses and lichens) was very slow compared with the rapid increase of the herd size, in the absence of any predators other than the regulated hunting by humans. He guessed that the island might have been able to sustain a long-term population of as many as 600 animals, but when it reached 2000, the destruction they caused to the productive sources of their own food was so severe that it led to a multiyear famine, driving the herd to the edge of extinction.

Many ecologists and population scientists have regarded this case as a chilling cautionary tale for our species, and rightly so. Certainly, the most immediate cause of our extraordinary population growth in the past 600 generations is the development of agriculture, which has provided us with more and more abundant food. Like the caribou on their island, we have been exposed to unnaturally rich living conditions, and we've responded with a population explosion – one that's not over yet. If we don't crash catastrophically in the next few generations, due to massive famines and so forth, then we'll be the first recorded case of a species that did *not* do so under those conditions. As I mentioned in the Introduction, a human die-off of this sort is just what Thomas Malthus predicted in 1798, and also Paul Ehrlich in 1968.

And yet, very few of the scientists who work on human populations still believe such a thing is likely. Since the 1970s, all the indicators suggest that we have dodged that bullet. Birth rates are falling, and falling fast. Food supply hasn't been strained to the breaking point... just the opposite. Per capita food supply has gone up dramatically in the past two generations, all over the world. The UN estimates that 11.3% of people were undernourished in 2014, down from 18.7% in 1992. That's an extraordinary improvement to come in just one generation, right at the height of an unprecedented global population explosion. Our innovations in the production and distribution of food have been exploding even faster than we have.

I'm certainly not saying that hunger is no longer a problem in our human world – it is. The fact that one out of nine people is still suffering from hunger is a nightmare almost beyond human conception. But even though that's true, in terms of evolutionary ecology, the want of food is not creating enough resistance to stop our population growth. We may not collectively have enough food (or food distribution) to let us all be healthy and happy, but we have enough to keep our population explosion going for another generation or more, if that were what we wanted to do. Consider this: Our population has been rising without cease for at least thirty generations, since the Black Plague – and yet, the percentage of humankind that is starving is lower today than in *any* of those previous thirty generations. In fact, we can safely presume that it's lower now than in any generation of humans, ever.

So nature's typical methods of regulating population growth don't have much of a grip on us. We're not only outside the reach of Darwinian control, but also (for the time being) Malthusian control as well. Under such circumstances, a naive scientist would boldly predict that our birth rates should be at an all-time high. And yet, as we've seen, they're actually at an all-time low, and falling fast. In any other population in the animal kingdom, if we observed a plummeting fertility level and wanted to prevent extinction, we'd start by providing protection and food to the breeding population. But in humans, those are precisely the conditions that have caused our birth rates to fall.

So, no, ecological pressures will not prevent satisfaction sterility from getting more and more severe. Like our talk of natural selection and the 'good of the species', this route of discussion has proved to be a dead end. What are we left with, then?

Any readers who have some expertise in human population studies have been waiting impatiently throughout this whole book for me to mention the words: demographic transition. Human population studies make up a big field, and experts of that field would never expect our species to be saved by evolution's mechanisms of natural selection, nor by ecology's mechanisms of resource limitation, nor by the pleasant myth of some hidden drive toward the 'good of the species'. In human population studies, the orthodox belief is that a population regulating mechanism exists as part of the process of socioeconomic development, and it is called demographic transition. Unfortunately, as we're about to see, there is no such thing.

Demographic transition is a pattern observed in population data, gathered from a number of nations around the world. As a nation goes through the process of development, it experiences a rise in per capita GDP, an accumulation of useful innovations, and many other signs of social improvement and economic growth. At some point in that progression, the death rate begins to fall, finally coming to a new stable level that is much lower than the original value. About two generations later, the birth rate also begins to fall. Some examples are shown graphically in Appendix Two, Topic D.

There are two very good reasons to believe that this pattern of 'demographic transition' is really two entirely separate phenomena, which should not be heaped together under a single name. The first reason is that there is no evidence of any cause-and-effect link between falling death rates and falling birth rates. The second reason is that there is no known, plausible way to explain any such link, which makes it unlikely that the link could really exist.

To see why this is so, let's try our best to make something useful out of the idea of demographic transition. To make the idea useful, rather than just a meaningless pattern seen on certain graphs, we have to claim that there's some sort of link between falling death rates and falling birth rates. Any such link would be great news, because it would imply that *some* sort of regulatory mechanism is at work, keeping our human populations trim and healthy. We could say: "Sure, falling death rates might cause a nation to go through a terrible population explosion, but two generations later, the birth rate slows down and everything levels out."

The actual data support everything in that claim, except for the last three words. Those three words can only be true if something is regulating the birth rate, keeping it in line with the death rate. What would that linkage be? Unless we're going to explicitly call upon supernatural explanations, the supposed chain of cause and effect would have to be this: the falling death rate leads to crowding, and the crowding leads to a lower birth rate after a couple of generations. When we use the term 'demographic transition', we are implying that humans limit their reproduction to prevent excessive crowding.

The obsolete idea that many species, including humans, self-regulate their populations to prevent excessive crowding was put forward by Scottish biologist V. C. Wynne-Edwards in 1962. It ruined his career. Wynne-Edwards felt that if you didn't believe in the self-regulation of populations, then you had to believe that the main thing that limits crowding is food shortage. He pointed out that, if that were so, then almost every animal in nature should be on the verge of starvation all the time. Since they're not, he felt that each species must have a fairly inflexible degree of tolerance for crowding. In his view, each individual in almost every animal species instinctively defends a particular size of territory, without regard to the amount of resources that the territory contains. He further believed that the reason that this sort of social system evolves is because animals that voluntarily limit their population density survive better, as a group, than animals that endlessly fight over resources.

But what mechanisms would allow a population to regulate its own density, according to Wynne-Edwards? This is an important question for us, because we need to identify some such mechanism in human populations, if we want to salvage the concept of demographic transition. Wynne-Edwards said that lots of species show their own version of demographic transition: according to him, natural populations respond to crowding by lowering their own birth rate.

Wynne-Edwards catalogued a remarkable number of population-limiting behaviors among various species, such as stress-induced infertility, and staying home with parents to care for younger siblings rather than seeking a mate. He proposed that such behaviors increase with crowding, serving as a dynamic regulator of population density. He extended these principles of population regulation to human societies, citing Carr-Saunders's 1922 writings on supposed population regulation systems seen in "primitive races" (to use Wynne-Edwards's term), through mechanisms such as infanticide and cannibalism. Of course, a lot of this stuff sounds pretty insane today, but at least actual mechanisms were being discussed. The term 'demographic transition', by contrast, has been around for decades and so far no one has proposed any plausible mechanism at all.

During the ten or twenty years that followed the publication of Wynne-Edwards's views, the scientific community roundly debunked him. Although his book contained a huge number of examples to support his idea, an even greater number of counterexamples was produced by his critics, leading to the opposite conclusion. During those same years, theorists struggled to come up with any scenario in which natural selection might operate on a group of unrelated organisms that were all competing for resources, and yet lead to the kind of population-regulating mechanisms Wynne-Edwards was proposing. None of those models ever stood up to mathematical scrutiny. Furthermore, nowadays a lot of the most convincing examples in Wynne-Edwards's book are easily identified as misconstrued cases of kin selection. Unfortunately, he had never heard of kin selection – that idea was first proposed two years after he published. By 1990, pretty much every evolutionary ecologist had concluded that strong, non-kin group selection is impossible. Modern ecologists feel confident that population densities are controlled by predators, parasite load, disease, and competition over resources, and not by adaptive mechanisms to forego reproductive opportunities for the good of the group.

The reader may be wondering by now what any of this has to do with demographic transition. Demographic transition is the last remaining major idea, in modern society, that is based on Wynne-Edwards's ideas. If the term is understood to mean an intrinsic regulatory mechanism that causes birth rates to fall in order to limit crowding in developed nations, then it is *pure* Wynne-Edwards.

So what's wrong with that? Why isn't it plausible to say, "Humans have the good sense to slow down their breeding behavior when their environment becomes rich with resources, in order to prevent over-crowding?" That notion falls apart, like all other Wynne-Edwards-style arguments, due to a problem called "the tragedy of the commons." The problem is this: If many people share a common resource, and no external force or authority is regulating the people's behavior, then each person can get a personal advantage by taking more than his or her share. That leads to competition, which depletes and eventually ruins the resource. In such a case, competition is bad for the group, but good for the most ruthless individuals.

In regard to human population growth, excessive crowding can certainly lead to the depletion and ruin of common resources. If some parents choose to have a lot of children despite the threat of over-crowding, then that might be bad for their community as a whole – but it still might be good for *them*. Exactly the same argument applies in a community that's suffering because too *few* children are being born... we can still count on individual families to produce the number of children they prefer, rather than the number that's ideal for the community. Again, all of this assumes that the parents aren't under the command of some regulating authority. We'll have a look at that possibility in the next chapter.

Once famine and the other blights of poverty have been removed – in other words, once a society is past the post-Darwinian threshold – there's no regulatory relationship in which crowding causes the voluntary fertility rate to go down. That means that demographic transition doesn't exist as a coherent phenomenon. In order for it to be coherent, demographic transition has to work in a chain of events, linked together as cause and effect. The chain should go like this: rising wealth causes lowered death rate which causes crowding which causes lower fertility. The last link in that chain, between crowding and low fertility, may not exist at all, and if it does then it's trivial in comparison with the *direct* link between rising wealth and falling fertility, which is satisfaction sterility. It's easy to find examples of poor, crowded nations with high fertility (for example, Nigeria), and rich, sparsely populated nations with low fertility (for example, the US). That suggests that human fertility doesn't fall due to crowding; rather, it falls due to wealth. So there is no regulating link between falling death rates and falling birth rates. They're two separate things.

The most straightforward explanation for so-called 'demographic transition' is that there are two cause-and-effect pathways, not one: A) Rising prosperity causes lowered death rates. B) Rising prosperity *independently* causes lowered birth rates. The mechanism for the first relationship is improved sanitation, medicine, food supply, etc. The mechanism for the second relationship is satisfaction sterility. Crowding has little or no tendency to lower birth rates (unless it creates a major famine), and this is true even if it continues for many generations. The two phenomena, falling death rate and falling birth rate, may often happen a couple of generations apart, but they are not linked in any meaningful way, and neither of them offers any mechanism of regulation to population growth or shrinkage.

So much for demographic transition. We've now eliminated most of the major candidates for the natural regulation of satisfaction sterility. Both natural selection and ecological stress have proved to be impotent, and both demographic transition and the drive toward the good of the species have turned out to be myths. Let me propose another potential regulatory mechanism – one that is tailor-made to help put a limit on satisfaction sterility. I don't think anyone has proposed this mechanism yet, but I imagine that someday someone will, so I'm going to prop it up as a paper tiger, just to shoot it down.

So-called 'birth dearth' is a growing concern in economics, and shows up more and more often in both the political and financial sections of the news. Birth dearth is a catchy phrase that summarizes a worrisome trend. In many regions of the world, the shrinking of labor pools threatens to cause a collapse in productivity, leading to economic recession. But perhaps there's a silver lining to this looming cloud. After all, if per capita GDP falls in the future as a result of low birth rates, then that might create a perfect counterbalance to satisfaction sterility – a made-to-order regulatory mechanism. Picture the beautiful balance of it: our prosperity rises and causes us to lose interest in raising children... so our population shrinks, causing our prosperity to slip back a notch or two. Soon, the whole system finds a point of equilibrium, and we all live happily ever after.

Of all the natural regulatory mechanisms we've considered in this chapter, this one is the most plausible. But is birth dearth real? Remember, by 'birth dearth', we don't just mean a fertility crisis, we mean economic recession *caused* by a fertility crisis, and falling birth rates. So far, there's no evidence to show such a trend in any nation – rather, there are only the grim warnings of pessimistic economists (and the taunting replies of optimistic economists). Let's consider the matter: how effective is falling fertility going to be as an engine for creating economic recession? Can we really count on future recessions to slam our economies hard enough to prevent satisfaction sterility, and save our species from extinction? That's a funny way to put it, I know, but there it is.

The notion of economic birth dearth has been around for an amazingly long time. The economist Joseph Schumpeter mocked alarmists in his own time (the 1940s) for warning that falling birth rates might cause recession. That might strike modern readers as an odd thing for people to have worried about back then, since the US population and the world population were both skyrocketing, and both of them have more than doubled since then. Schumpeter's point, though, was that even if population growth *did* slow down, economies wouldn't 'flop' (as economists used to say back then) because innovations would come along and enhance per-worker productivity. As we'll see, it turned out that he was far more correct than even he would have believed possible.

Birth dearth is only a vague issue in US political discussion at the time of this writing, but it's already a very immediate worry in many nations. One strong example is Japan, where birth dearth threatens the national pension scheme (roughly the equivalent of the Social Security system in the US). When Americans worry that they won't receive Social Security benefits in their old age, they are usually worried that the economy will fail due to the national debt, or crises in the finance sector, or other systemic failures. When the Japanese worry that they won't receive their government pensions, it's usually because they doubt that enough young Japanese workers will exist to support all the old people. That's birth dearth, in its purest form.

Japan's worries about their pension scheme are well-founded. Theirs is one of the fastest-shrinking nations in the world, and also has one of the most rapidly aging populations, due to increasing life expectancy. But there is also a special, regional consideration in Japan, namely, that nation's reluctance to allow the mass immigration of foreign workers. In Europe, where the native population's fertility rates are plummeting, there is a powerful influx of workers from the Middle East and North Africa to fill the economic void. In the US, a similar influx moves constantly north from Latin America. Japan is in a geographic position to receive an endless supply of new citizens and residents from other nations, especially the Philippines, but it resists this geopolitical migration pressure. Regardless of whether the reader approves or disapproves of that policy, it exists, and is sufficient to explain why birth dearth is feared more acutely in Japan than in other developed nations.

Opening borders to foreign workers can certainly relieve birth dearth in the short run, but it's not a cure. What if the entire world becomes prosperous in the future, and even today's poorest nations become wealthy enough to experience satisfaction sterility? That may sound far-fetched, but a century ago, almost no one would believe how wealthy and high-tech the world was going to become by our time. So let's stretch the limits of our optimistic imaginations, and picture a future world where there really aren't any poor nations, at least by current standards. A side effect of that worldwide prosperity will be global disinterest in having big families. The huge international labor markets that currently exist will dry up, and the richest nations will no longer be able to tempt armies of workers to trudge across their borders and do manual labor. Will their economies 'flop'?

The data suggest not. As Schumpeter predicted, the world has seen a steady and powerful trend toward innovative increases in productivity per worker. This trend, often nicknamed 'automation', reduces the number of workers needed to get a particular job done. Sometimes this really does involve automation, in the literal sense of augmenting or replacing human workers with machines. Sometimes it just involves improved techniques, materials, and so forth, rather than the addition of new machines to the workplace. But the result is the same: rising productivity per worker. That can result in improved output, or people getting laid off, or both. Regardless, automation (in this broad sense) works directly against birth dearth. Every advance in automation makes it a little less likely that a shrinking labor pool is going to cause another recession.

Let's be sure we see this point clearly, since it's such a key factor in predicting the future outcome of satisfaction sterility. Consider two different ways of doubling worker productivity: 1) I bring a new machine into a factory, and it allows a single worker to do the labor that two workers did before. 2) I introduce a new fertilizer to a farm, and it allows one acre of crops to yield twice as much food under the same labor practices as before. The first of these two cases is automation in the strictest sense of the word, while the second case is not. And yet, both have the same effect upon the relationship between laborers and productivity: one worker can do a job that used to require two. For the rest of this chapter, the term 'automation' will refer to *any* innovation that reduces the number of man-hours needed to achieve a certain amount of production in a given time.

Now, when we say that automation can permanently remove the problem of birth dearth, that doesn't mean that populations are going to stop shrinking. It only means that the economic problems associated with shrinking population will go away. It means that per capita GDPs of nations may go ahead and keep rising in the future, even if labor pools keep getting smaller. That assertion might sound like wild-eyed optimism to some readers, but only if they're forgetting our world's economic history over the past century or two. For a long time now, automation has been decreasing the amount of labor needed to achieve a given amount of productivity. It has been doing so at a staggering rate, which is a lot of the reason our modern world is so different from the world of, say, 1800.

Consider this fact: the average American worker in 2013 could generate as much real value in a workday as 3.8 American workers in 1947. An increase of nearly fourfold, in just three generations! And remember, even at the start of that period, in 1947, the US was already the most productive nation in history. The US came out of World War II sporting military and industrial sectors that were so awesome that just half a generation later, the president was actually warning the American people to watch out for them.

A statistic like, "One worker in 2013 could do the labor of 3.8 in 1947," is nearly incredible, and a skeptic would be well within his or her rights to try to challenge it on technical grounds. For example, a lot of the hands-on labor of American factory work was gradually sent overseas during that period. Could that have skewed the statistics? Perhaps all of that global-ization made it possible for multinational corporations to show profits in a wealthy nation like the US, while the actual, physical productivity was occurring overseas in a cheaper labor pool. Let's consider that possibility for a moment.

Almost all statistics involving the economics of globalization are contentious. Nonetheless, the basic assertion that per-worker productivity is rising, and rising fast, is not easy to shake. In particular, agriculture gives strong support to the claim, because it shows the 'automation' trend strongly, despite the fact that very little of US agricultural productivity has been outsourced overseas. A lot of American shoes and cars may be assembled abroad nowadays, but most American corn and beef are still grown domestically. During the period from 1948 to 1989 in the US, the productive output per farm worker (including estimated numbers of undocumented immigrant workers) more than quadrupled. The average American farm worker in 1989, legal or undocumented, could produce as much real value in food products in a workday as 4.1 American farm workers in 1948. It's almost impossible to account for that in any way except through automation. The per-worker productivity has soared since 1948 due to the adoption of innovations that rendered a great deal of human labor obsolete, making it possible to achieve higher worker efficiency. In the case of farming, those innovations included improve-ments in crop genetics, irrigation, and pesticides, to name a few.

The constant accumulation of embedded innovations in the workplace – whether office, factory or field – is making human labor a smaller and smaller component of productivity. That's certainly been true for decades, probably for centuries, and is more true now than ever before. Will the rate of automation outstrip the shrinkage of labor pools, making birth dearth irrelevant? The statistics strongly argue that it will. It's hard to see any rational way to support the counterargument, which would be: "We're going to run out of innovations and hit an impenetrable ceiling of per-worker productivity." After all, what would *prevent* future technologies from replacing yet more people with machines and other innovations? In fact, that pessimistic argument sounds very similar to the warning given by

Malthus, over 200 years ago, when he said that innovations in agriculture couldn't possibly keep up with rising population. As we've seen, he was dramatically and demonstrably wrong.

Some technologically conservative readers may feel that it's unrealistic to imagine automation changing those aspects of the workplace that require the human touch. A critic might argue: "In factories and farms, that sort of thing might go on, but surely not closer to home." This view is misconceived. In fact, the sector of the labor pool that traditionally handled the jobs that required the *most* human touch has already been replaced with automation in the US, almost entirely. That sector consisted of domestic servants. Servants were once a universal aspect of middle class home life, but the occupation of servant is nearly extinct in American middle class homes today. Imagine telling a middle class family, five generations ago (back in 1900) that their descendants today would often live in houses with no servants whatsoever, and yet have lives that were even more comfortable than theirs. Not many people back then would have believed it possible. So what happened to all those servants? They weren't replaced with robots. They were replaced with dishwashers, microwave ovens, electric water heaters, vacuum cleaners, televisions to nanny the kids, and so on.

Given the relentless accumulation of innovations, no long-term harm will come to the economies of nations, no matter how much their populations shrink. Automation trends will make workers obsolete faster than satisfaction sterility can reduce their numbers. That means that, in general, there's not going to be a global birth dearth crisis – no worldwide plague of recessions caused by falling population. The fertility crisis is permanent and real, but economic birth dearth worries are a passing thing.

Regional exceptions will occur, of course. I've already mentioned Japan, which is both shrinking and aging so quickly that they are likely to be slammed by a birth-dearth recession in coming decades, unless they relax their immigration policies. But if that recession does come, it will pass in good time. An international observer might worry that *some* countries will fail to automate quickly enough to prevent long-term birth dearth... but I wouldn't worry about Japan. Since the late Meiji Period, Japan has been among the forefront of nations in solving problems through innovative automation. It's a pretty safe guess that they're ultimately going to solve their birth dearth problem that way... not by making more kids, but by making workers obsolete.

In general, birth dearth (like satisfaction sterility) is a problem that only strikes developed, wealthy nations. But there are exceptions. Currently, the most important set of examples is seen in the nations that once made up the Warsaw Pact nations, known in the Cold War days as the Eastern Bloc. Over half of those Eastern European and ex-Soviet nations are currently held back from economic growth by (among other things) very low birthrates. Effectively, these are developing nations that are suffering from birth dearth. That's odd, since birth dearth is usually described as a blight threatening the wealthiest nations. It's important to notice, however, that the afflicted nations in this group share a strong historical factor: each of them had a Soviet-style communist economy that vanished one generation ago. Their form of economic blight, characterized by low birth rate, is not seen in other regions with similar per capita GDPs. The most common belief among international observers is that the Eastern Bloc's regional malaise is an exception, not a rule, and can be expected neither to spread, nor to persist for very many generations.

But *could* birth dearth in developing, ex-Communist nations like Romania and Albania prevent "modernization" (especially in the form of automated productivity), and create a long-term block against economic resurgence? Here's the reason it's worth asking that question. An economy that permanently stalls due to birth dearth is in a regulated state that halts the progression of satisfaction sterility. That's what we've spent this whole chapter looking for. Imagine it: Albania gets a little richer during some future period, and as a result, satisfaction sterility lowers the nation's birth rate. That causes a decade or so of recession, due to birth dearth. The economic hardships increase the personal dissatisfaction of the citizens, and satisfaction sterility starts to go away. Birth rates go back up, relieving the birth dearth problem and saving the economy. The result: another rich decade begins, and the cycle repeats forever. Problem solved! Maybe the rest of us will go extinct, but not Albania.

One trouble with that notion is that Albania and the other ex-Eastern Bloc nations don't exist in an international vacuum. Nations trade and otherwise interact with other countries. Almost any developing nation can absorb automation techniques from richer neighbors without having to 'evolve' to a point where improved automation and other productivity enhancements appear spontaneously within their own infrastructure. The nations of Eastern Europe and the ex-Soviet republics all trade heavily with the European Union, serving their richer neighbors as a source of cheap labor and raw materials. It strains belief to think that that's going to go on for the next two or three generations without a great deal of automation and other improvements in per-worker productivity being absorbed by the region, in addition to whatever they come up with on their own.

One more point bears mention here. In Chapter Four, in the story of the Three Brewers, I mentioned the importance of maintaining both productive infrastructure *and* military readiness, if an autonomous society hopes to remain stable for long. Perhaps the oldest and still one of the most crucial military resources a nation can have is its armed forces, both standing and in reserve. This is a 'labor pool' of trained soldiers who can be called upon to carry out a military action, as needed. Tiny nations with tiny armies are easily overwhelmed by nations with immense armies, as when the US invaded Grenada in 1983. So we can imagine a military version of birth dearth, in analogy to the economic version we've been talking about. If falling birth rates cause a nation's standing and reserve armies to shrink, perhaps the nation's military preparedness could collapse. This would be the military equivalent of a birth dearth recession.

As in the economic case, however, it's very far-fetched to imagine that this will be a major factor in future world history. The numerical size of the standing and reserve armies is only one part of a nation's military strength. The quantity and innovative quality of its armaments is of greater importance, and the relative importance of this factor (compared with manpower) continues to rise and rise. This has been true since the dawn of colonialism, when forces armed with guns first invaded lands protected by larger armies without guns. Famous examples illustrate the point: In the Battle of Cajamarca in 1532, 168 Spanish soldiers with guns defeated about

6000 Inca warriors wielding clubs. Sixteen generations later, at the Battle of Rorke's Drift in 1879, about 150 British soldiers with guns defeated about 4000 Zulu warriors wielding spears.

The innovation of the gun allowed one soldier to act as effectively as several soldiers had done before. It fits neatly into our working definition of 'automation'. Just as a new fertilizer might make a farmer more productive, or a new machine might replace a factory worker, so a new piece of military technology can augment manpower in the armed forces. Historically, this process went through a rapid lurch forward during World War I, with the introduction of the warplane, the tank, chemical weaponry, and the first systematic use of machine guns. In recent decades, a new lurch forward is occurring. Drones have been replacing manned aircraft in the armed forces of many developed nations, such as the US. Other forms of remote-controlled or even fully robotic combat machine are under rapid development, and their deployment is increasing in both frequency and range of purpose.

But perhaps the most conspicuous example of military 'automation', in the broad sense of replacing human soldiers with innovative alternative threats, is the nuclear bomb. It only takes one person's finger to launch a nuclear missile attack, once the systems have been put in place. Obviously, it takes more people than that to establish and maintain those systems in readiness, but if we take a headcount of all those people and compare their number to the threat created by their weapons system, we still find that they are easily the most daunting sort of military group in history. The threat-to-manpower ratio of a working nuclear arms program is far greater than that of any conventional army. Furthermore, any nation, no matter how small and otherwise ill-equipped, that gets possession of The Bomb increases its military credibility immensely. That's true even if that nation's population and the headcount of its army are decreasing.

So, innovations increase the effectiveness of both productivity and armed forces. Those increases are realized continually and at a very steep rate. Innovation will eventually release any economy from a recession caused by birth dearth, because it will cause a nation's per capita GDP to grow even if its labor pool is shrinking. Decreasing birth rates are not going to impede economic growth of either developed or developing nations in the long run, because labor pools will be replaced with automation faster than they can shrink.

Those opinions seem plenty sound, and they're based on solid evidence... but let's not put any weight on mere opinion. What if those opinions are wrong? That is, what if birth dearth proceeds faster than the economic improvements of innovation? Even in that unlikely event, as it turns out, birth dearth won't stop satisfaction sterility for long. Here's why.

If we imagine a nation with plummeting population size, sluggish innovation, and no substantial source of immigration, then we can safely predict a long, bitter recession. That will increase everyone's dissatisfaction, certainly, and in turn that's going to relieve their satisfaction sterility. Fertility will start to rise a bit, and something like a stable situation will ensue. But in that stability, one thing will *not* be at a standstill: innovation. New ideas and machines and techniques will keep appearing, regardless of whether that happens slowly or quickly. No matter how slowly the accumulation occurs, eventually the innovations are going to pile up. By their very nature, the innovations are intended to decrease dissatisfaction, to make things work better, and to alleviate some of the human misery. As that goes on, prosperity will gradually rise, and then satisfaction sterility will set in, and fertility will again begin to fall.

So, birth dearth simply doesn't have the potential to stop satisfaction sterility. In fact, as we've seen in this chapter, *no* natural process can do so—economic, sociological, or biological. So if anything is going to stop the shrinking of the human population, it will have to be us. Let's roll up our sleeves and start looking for a solution.

9

ARTIFICIAL PROTECTION

Parts I and II of this book introduced the concept of satisfaction sterility, a trend that causes whole populations of people to want fewer children. Then, in Chapter Eight, we found that there's no natural barrier to prevent that from happening. If our descendents really all decide to stop reproducing, then neither biology nor economics is going to stand in their way. Like a ball rolling downhill, our global fertility rate will just keep getting lower and lower, until no one has any children at all.

If the reader has been following this argument closely, then he or she might reasonably be thinking: "Big deal. We can fix *that*. Should be easy!" Even if this book's argument has been clear and convincing, there's no reason to think that satisfaction sterility is a matter of real consequence – not from what we've seen so far. Our species has a long history of causing its own, artificial problems, and then facing them down with technological and legal solutions. We were destroying the stratospheric ozone layer in the 1970s, so most countries agreed to ban the use of chlorofluorocarbons in spray cans and air conditioners. We seemed on the verge of destroying most of the world's large cities with nuclear bombs during most of the late 20th century, but we managed to put the Cold War behind us without actually doing it. So if our population someday really gets to the point where it's shrinking toward extinction, surely we're not helpless... we can *do* something about it. We've put men on the moon; we've driven smallpox extinct. This should be an easy one.

But putting an end to satisfaction sterility turns out not to be quite so simple as that. True, if the question is, "What can we do to prevent satisfaction sterility?" then the short-term answer to that question is: "We can do *lots* of things!" This chapter will have a look at our several options, one by one. But as we do so, it's going to become clear that the long-term answer to that same question is: "Nothing we do is going to improve matters at all."

As modern people, that idea of helplessness rubs us the wrong way. Our species has an amazing track record of achieving whatever we set our minds to. The problem, in this case, is that we can't really set our minds to doing something we don't want to do. Worse than that: we can't make our future descendants change their minds about anything, no matter how badly we who are alive today might want them to do so.

Let's start by having a look at our options, so we can see where the impediment lies. The solutions to satisfaction sterility described in this chapter are 'top-down' solutions. By that term, I mean that the motivation to get the job done doesn't appear spontaneously among us citizens, but rather comes down to us from those whom we recognize as authorities. Top-down solutions are ways to limit satisfaction sterility that are imparted to the general population by bodies of authority.

There are four routes to a top-down solution to satisfaction sterility. These are: legal, religious, ideological and technological. Let's see if we can find a viable solution using one or more of these. We'll start with the most straightforward means of changing the habits of a population: the law.

Probably no one feels good about the idea of passing a law that would simply order citizens to produce children against their wills. But the force of law doesn't have to be applied so bluntly and crudely as that. It's possible, in principle, for governments to sanction citizens who don't produce their quota of children, without crossing any absolute ethical boundaries. Those sanctions could be strong or weak, and needn't be as cruel or intrusive as jail sentences. Withholding tax breaks or other government-controlled perquisites can be very effective in prodding a population to change their collective habits. For that matter, there's an even simpler method of legislating fertility, which is to prohibit contraception. In the past, contraception has usually been limited by law due to religious influence, when it's been limited at all. But if population shrinkage becomes a matter of government concern, a nation's legislature can boost the country's fertility rate by exactly the same sort of prohibitive law.

Figure 8. The era of population enhancement programs has begun.

This billboard outside Tehran is part of Iran's official program to increase birth rates among its populace. The program was launched in 2014, and involves both legal measures and propaganda, such as the image shown here.

The world's first fertility-enhancement law was proposed very recently. In August 2014, Iran's parliament responded to the country's low fertility rate (1.85 children per woman) by overwhelmingly passing Bill 446, entitled: *Bill to Increase Fertility Rates and Prevent Population Decline.* The bill was awaiting approval by Iran's Guardian Council at the time of this writing. If it becomes law, this bill will render illegal some of the most widespread methods of family planning in Iran. Legislation against family planning is familiar enough to Western readers, of course. The thing that's unusual about Iran's Bill 446 is its title, which makes it clear that the purpose of the law is not religious, but rather to combat voluntary infertility... thus shifting reproductive rights away from individuals and into the hands of state authorities. In fact, the bill is part of a general program on the part of the government to raise the birth rate (see Figure 8) .

Although government programs to legislate increased birth rates are still rare, there have been some conspicuous laws that have been passed in recent history, intended to *lower* fertility. Those of India and China are the best known examples, and are also the ones that have been implemented on the largest scale. It's worth having a quick glance at these cases, because fertility-increase laws in the near future may experience similar challenges to those that have faced the fertility-decrease laws in India and China.

India has been encouraging citizens to undergo surgical sterilization since 1976, as part of a legally mandated policy to decrease the population growth rate. Given the comparative reputations of China and India with regard to international transparency, it's odd to note that vastly more contention and disagreement exist with regard to India's policies in this matter than China's. India continues to officially encourage, or perhaps coerce or even force, a large number of women to undergo tubal ligations annually... perhaps millions of women per year. Both the numbers and the circumstances of these operations are hotly contested on all sides, and I make no claim to know the truth of the matter.

By comparison, the one-child policy in China, under implementation since 1980, has been more of a success, both in terms of effect and in terms of international approval. The punishment in China for exceeding the household quota of children is generally just a fine, and many urban households are wealthy enough to pay the fine and go ahead and have extra children if they choose. Despite all that, fertility rates in China are far below the replacement rate – about 1.55 children per woman in 2014. Note that the population of China is still growing, but this is not a sign that the policy has failed. As discussed in Chapter Seven, it merely means that increasing economic prosperity is causing a boost in longevity, so that death rates are even lower than the birth rate, temporarily. People are living longer, but they won't live forever. In another decade or so, the death rate will go back up to its normal, sustainable level, and China's population will begin to collapse at an extraordinary rate.

China's government is keenly aware that their nation has now joined the ranks of nations that are experiencing strong, voluntary infertility. They may not be as terrified of a birth-dearth recession as is (say) Japan, but that specter is starting to loom in their future. Given the relative success of China's fertility-reduction laws, are they going to be able to use legal pressure to enhance fertility someday? They've already made a first, tentative effort in that direction, but it went very poorly, to everyone's surprise. Here's what happened:

In 2014, China relaxed its one-child policy, offering a new loophole to let some couples have a second child. But by the end of the year, the number of eligible couples who had applied for permits was less than half what the government had expected. The widespread belief had been that Chinese couples *wanted* more children than their laws allowed... but that belief turned out to be false. So why don't Chinese parents want more children? In a survey, over 75% of respondents said that they didn't want a second child because they couldn't afford one. This is exactly the situation that has existed for a generation now in Japan, as we saw in Chapter Eight. It's absurd to believe that the richest generation in Chinese history "can't afford" to have as many children as any previous generation. A more rational way to interpret that claim is that this generation simply doesn't want as many children as did their ancestors. They're experiencing satisfaction sterility.

Perhaps China, Iran and many other countries will eventually pass laws offering financial incentives (tax breaks, fines, etc.) to pressure the populace to reproduce more. Perhaps these will occur side-by-side with partial or even complete bans on contraception and family planning techniques. Who knows, perhaps some authoritarian governments may someday even threaten parents with jail if they fail to produce their quota of kids... but that sounds unlikely. Physical infertility problems are common enough that almost anyone can plead: "We're trying as hard as we can, officer!" So fines and contraception bans are likely to be the main legal levers that are used to boost the fertility of nations, in coming years. Will they work?

In the short run, such laws will surely cause more children to be born. The long run is a different story. The enforcers of fertility-boosting laws will have to struggle against the ever-growing inertia created by a progressive tendency toward voluntary infertility among the citizens. In the first generation after such laws are passed, a lot of couples might willingly opt to have just one more child, in order to improve their tax status, avoid a fine, and maybe buy a new car. Many couples around the world are sitting on the fence, so to speak, unable to decide whether one kid was enough, or if maybe they should have a second. A little financial encouragement from the government will nudge them to take the plunge.

But after another generation has gone by, satisfaction sterility is going to be *much* stronger than it is today, and such laws will stop working. Almost no one will be willing to have an extra child... they'll prefer to pay the fine. If that statement sounds like a crystal-ball prediction of the future, it's not. Satisfaction sterility has been increasing globally at a steep rate over the past few generations, and we've already seen that no natural resistance exists to keep that from continuing into the future. Furthermore, there's already a large proportion of couples in developed nations who would refuse to have an extra child, even if that meant paying a fine. Obviously, as satisfaction sterility gets stronger, that proportion is going to grow.

Banning contraceptives is even more short-term in its effects. True, in a developing nation, a ban on family planning will probably increase the birth rate by creating hordes of unwanted children. But as that developing nation's economy gets stronger, the law will soon have little effect except to create a lucrative black market for condoms and birth-control pills. Long before that, most citizens will make adjustments to the prohibitive laws, and figure out ways to live their normal lives without too much risk of unwanted pregnancy. The main effect of laws like the one proposed in Iran's Bill 446 are to make it a little easier for a man to impregnate a woman against her will. But that's already legal, not only in Iran but in almost all nations. In the US, for example, there are no laws against a man putting holes in condoms, switching birth control pills with fakes, or otherwise tricking a woman into pregnancy. A large study in 2010 found that 8.6% of American women have experienced such efforts to get them pregnant against their will. Despite that, US fertility rates continue to fall.

So we have some reasons to doubt the long-term viability of laws intended to boost fertility. As a nation's citizens get richer over coming generations, their interest in having and raising children will get lower and lower, creating ever-greater resistance to such laws. But there's an even worse problem with legal solutions to the satisfaction sterility problem. As each generation passes, we must anticipate that the government's drive or will to continue the program will collapse even faster than the people's will to obey its laws. Let's see why that's the case.

There are three basic reasons for a legislator or judge to support a law: because it's good for the state, or because he or she feels the law is right, or for personal gain. The third of these is corruption, by definition. It's hard to imagine any way in which corruption could ever help prevent satisfaction sterility, so we'll concentrate on the other two.

A congressman might vote for a fertility-boosting law in order to help the nation, if he or she anticipated a recession due to birth dearth, or a weakening of the military due to dwindling recruitment. In Chapter Eight, we saw what will happen if such laws fail to pass. Improvements in automation will increase the productivity per worker (and the military power per soldier), until eventually the problem goes away. As workers and soldiers become ever smaller components of a society's productive and military sectors, falling population becomes less and less relevant to the nation's well being. That doesn't mean that legislators will ignore the issue of a vanishing population, but it does mean that they won't be inclined to vote for it as a way of strengthening the nation's economy or military. The only thing that might make them support such laws, in the long run, will be their personal feeling that it's the right thing to do.

It's tempting to imagine that lawmakers in the future will act to stop the progressive collapse of national populations as a matter of simple principle. But there are two strong reasons to doubt it. The first reason is the resistance that they'll encounter among their constituents. As prosperity keeps rising, it's not only going to decrease the interest that we citizens have in producing children, it's also going to give us more and more political empowerment. In a nation where per capita wealth is high enough (and sufficiently well distributed), the opinion of the average person becomes a powerful force in the halls of government. In any nation composed mainly of rich people, it's going to be impossible to maintain draconian laws that bully the general public to make babies they don't want. Such laws might appear briefly, but can't exist for more than a generation or two, at the longest. Remember the Prohibition, when US citizens first found themselves rebelling *en masse* against the law of the land – by simply ignoring it. The word 'scofflaw' was invented during the US Prohibition, to describe this amazing new power of the citizenry to get their way by simply ignoring an unpopular law. It's no coincidence that this happened during America's first period of widespread prosperity, the Roaring Twenties.

The second reason that government officials will not be able to maintain legal pressure to force fecundity in their nations on grounds of principle is that they, personally, will lose interest in those principles. The members of a country's government are typically richer than the average person. That means that they'll experience more satisfaction sterility than the national average. So not only will the populace become less willing to cooperate with baby-making laws as time goes by, the government's will to continue such a program will collapse even faster than the people's will to obey. If automation has made it possible to grow the economy without growing the population, why would such leaders want to defend a bunch of unpopular baby-making laws?

To see this last point more clearly, let's imagine a simplified case. Picture for a moment a society that's perfectly split into two groups, the Haves and Have-nots... like some sort of Marxist nightmare. In such a society, of course, it's the Haves who will make the big legislative decisions, and their feelings will be most important in determining state policy. Time goes by, and innovations keep piling up, and the economy gets more and more prosperous. Although the Haves get more than the Have-nots, even the Have-nots gradually get richer. (Note that, if they don't, then the problem of a fertility crisis will never come up.) By the time the economy has grown so strong that the Have-nots begin losing the desire to raise children, the Haves are already way, way beyond that point. The population begins to fall, but the question of legislating increased fertility isn't of any interest to the Haves who are running the government. As long as there are machines to do the work and guard the borders, they could care less whether the Have-nots raise children. In fact, they're relieved to see the crowded parts of the cities finally thinning out a bit.

This two-tier model is an extreme and (hopefully) unrealistic picture of future societies, but the same argument applies to a society that is divided into an arbitrary number of social strata, rather than just two. With each step up the ladder of social classes, we find that the average person is a little richer, has a little more say in government and the management of society, and is a little more strongly affected by satisfaction sterility. So the fact that wealth and class are distributed continuously in a society doesn't make the government any more likely to offer resistance to the general trend of satisfaction sterility – not in the long run.

We're zeroing in on the core issue, now. The problem isn't one of lazy authority and weak command; rather, it's a progressing change in the way people think. A more direct attack on the problem would involve social changes that are aimed at altering the value systems by which we average citizens live our lives. Perhaps, if that's the case, the way to ensure public support for population-enhancement programs is through propaganda and other indoctrination techniques. Certainly, those sorts of approaches can sometimes be very effective even without any legal apparatus to enforce the policies.

Systems of thought that are intended to change the views and behaviors of large groups of people are traditionally divided into two categories: religions and ideologies. Let's start, then, with religions. Can religious dogma and authority be used to coax a population of people to have more children? They certainly can! In fact, these methods have a proven track record, and one that's far more impressive than any government program, whether based on law or propaganda, at least so far. Perhaps the Catholic religion is the most familiar example of a successful religious intervention in population decline, at least for readers living in Western nations. Catholic communities and nations had a reputation for high fertility rates, right up until a generation or two ago.

That reputation was well justified. For a long time, many Catholic communities in wealthy countries managed to hold onto the ancient values that had once impelled most families, regardless of religion, to have as many children as possible. That outdated practice was common among Catholics until just two generations ago – even rich Catholics living in rich nations that had low birth rates overall. The trend was encouraged and reinforced by official church doctrine. To give a specific example: Catholics had much higher fertility than non-Catholics in the United States in the 1950s and 1960s, reaching a peak difference in the early 1960s, with 4.25 children per Catholic woman, as compared with 3.14 per non-Catholic woman.

But then, in the late 1960s, Catholic fertility in the rich nations began to decline. Vatican authorities responded by formalizing their fertility-enhancement policy, changing it from a vague set of encouragements to a concrete command. In 1968, Pope Paul VI issued the *Humanae Vitae*, a document that explicitly prohibited birth control. It said: "Excluded is any action which either before, at the moment of, or after sexual intercourse, is specifically intended to prevent procreation." His successors John Paul II and Benedict XVI referred to the *Humanae Vitae* as church law, and officially sustained the Catholic prohibition of all artificial birth control.

And yet, despite this direct, long-term exercise of religious authority, the difference in fertility between Catholic and non-Catholic families in the US effectively vanished by the mid 1970s and has never bounced back. The same trends have been seen in Europe, even in the most strongly Catholic nations. Poland and Italy, the two most Catholic large nations in the world (86% and 85%, respectively) are now also among those with the lowest fertility rates. From 1968 to 2013, despite strict religious instructions to avoid family planning, fertility in Poland fell from 2.24 to 1.26 children per woman, as the per capita GDP rose from $3200 to $13,700 in real terms. In Italy during the same period, per capita GDP rose from $13,000 to $35,700 in real terms, while fertility fell from 2.49 to 1.39.

Although Catholicism is a conspicuous example of religious authority intervening with human subpopulations to enhance their fecundity, this is also a common property of many other religions. For example, in some parts of the world, communities belonging to various Muslim sects show much higher fertility than non-Muslims living under similar socioeconomic conditions. Among the nations that were once members of the Eastern Bloc, those that are primarily Muslim have nearly twice the fertility rate found in the non-Muslim nations, when adjusted for the effects of wealth. That situation is examined more closely in Appendix 2, Topic F.

Many readers might be skeptical of the long-term viability of these religion-based social trends. As we've seen, the Catholic example of religious fertility enhancement has completely collapsed, at least in developed nations, despite all of the church's efforts to keep the pressure on. That's a sensible criticism, but religious dogma isn't the only basis for propaganda and other indoctrination techniques that can be used to prod a demographic group to make more children. Ideological doctrines can also do the job.

For example, the Third Reich ran a propaganda campaign intended to increase reproduction among certain ethnicities. The *Lebensborn* program was an international effort run by the SS, encouraging enhanced fertility within the supposed "Aryan" racial subpopulation of Europe. A propaganda program urged selected women to conceive embryos with SS soldiers and give birth anonymously. The babies were then either adopted by Nazi families, or raised in an international chain of SS-managed birth houses. Although the program was active for ten years, it may have produced as few as 8000 children – fewer than 0.1% of the German children born in those years.

A government can also try setting up a sort of state religion, jazzing up a dull system of ideological propaganda by giving it the glamour of religious dogma. This is seen in North Korea's semi-religious ideological system, known as *Juche*. Maybe some combination of religious and state propaganda could convince people to reproduce more than they are naturally inclined to do, especially if such a program is applied alongside tax breaks and other legal incentives. But this still sounds suspiciously weak and toothless, seen from a long-term perspective. After all, we've already seen that the legal incentives are likely to crumble away as time goes by. We've also seen that religious pressures, even when delivered as direct commands by the Pope

Figure 9: Fertility as duty to the state.

Political ideologies can be used to raise fertility rates, even without recourse to legislation or religious pressure. In Nazi Germany, the SS coaxed so-called "Aryan" women to have extramarital babies for the good of the Third Reich, in a propaganda campaign called *Lebensborn*. The SS raised many of the babies in birth houses like this one.

(the top authority in the world's biggest religion) have no effect at all when they are pitted against a force as powerful as satisfaction sterility. So why should we believe that the combination of the two is going to work any better?

There's another tool that could be added to this effort, one that might drastically increase the leverage of authorities when they try to raise fertility rates among us common people. In principle, if future laws allow it, the use of automation and factory techniques could help boost a nation's production of babies, just like it does with our production of chickens and cattle. A government that was really determined to enhance population might turn someday to such schemes, and begin to mass produce new citizens.

I think most of us are mighty uncomfortable even talking about this... but the technology already exists. Frozen sperm and egg supplies could be used to assemble embryos in industrial laboratories. Professional surrogate mothers already exist – women who are willing to hire out the use of a uterus for nine months to anyone with the cash to pay for it. And who knows? Maybe someday even that service can be replaced with a cheaper alternative... some kind of mechanical uterus, perhaps, created along the same lines as mechanical heart valves and hip joints. After each factory baby is born, very little individual care will really be necessary, not if a future regime has really made up its mind to mass produce citizens. Nurseries and boarding schools can be set up with a minimal adult-to-child ratio, and that ratio will fall quickly as the years go by and new technologies replace the human touch. Kids already get a lot of their nannying and education services through electronic displays of one sort or another. One day, maybe robots can handle the whole job of child-rearing, from start to finish.

Perhaps we've uncovered a really effective (if ugly) future scenario for putting an end to satisfaction sterility, and preventing human extinction. It sounds like all it will take is for some hard-line political regime in the future to get serious about the problem of dwindling population, and start building baby factories. The rulers of that regime might even perceive the biggest problem that looms in their own future: namely, the fact that their own grandchildren are surely going to lose interest in perpetuating the harsh but species-sustaining factory system. To head off that possibility, those iron-fisted rulers might very well leave behind a set of stern laws and a quasi-religious state ideology to foster the belief that producing children is a sacred duty, even long after everyone has forgotten why. Like Adolf Hitler, those far-sighted despots might confidently believe that their new order will last for a thousand years.

That certainly all adds up to a shocking science fiction story. And who knows, maybe something like that will really come about, in some nation, sometime in the future. We can laugh at the idea, but let's not forget that almost everything in our modern world would have sounded like science fiction to intelligent people a hundred years ago. A crazy scheme like that might even work... for a while.

But as we gaze into that creepy future vision, we find (for better or worse) that we eventually run into the same problems that we encountered when we were considering child-making laws. The better the system works, the more likely it will be that future generations are going to become too comfortable to really give a damn about forcing other people to reproduce. That's especially true if there's no economic, military or political motive for doing so. After two or three generations, all those stern laws will be re-written, all the fanatical ideology will become a joke, and the citizens will be busy enjoying their wealth in whatever way they choose. In the eyes of that young generation, the fanaticism of their grim-minded grandparents will seem kind of silly, if not insane, and the whole child-factory scheme will just fizzle out.

As people's internal drives to have and raise children keep falling, their willingness to follow religious or state encouragement toward fertility must inevitably erode away after a generation or two. Even the most effective propaganda machine can only sustain an unpopular trend for a limited number of generations. People will eventually do what they want to do. Wealth gives people opportunities for individual satisfaction that their ancestors didn't have, and this makes them lose interest in having and raising children, due to satisfaction sterility. When the average individual drive toward raising children becomes sufficiently low, people stop obeying orders to have them, regardless of who is giving those orders. The Catholic refusal to obey the *Humanae Vitae* is a huge historical example, but it's only the first. Let me confidently predict that we're going to see a lot more of that, as time goes by.

So, while it's true that citizens could be manufactured in automated baby factories, there would have to be someone around who felt there was some *reason* for doing so. A desperate scheme like that only makes economic sense if the mass-produced workers and soldiers are valuable to their manufacturers. As we've seen, the economic and military value of people is rapidly vanishing, when compared to the rising value of innovative technologies. If everyone is already getting everything they could possibly need and want from the highly mechanized infrastructure of their rapidly depopulating society, the last thing anyone will want is a baby factory. After all, any future citizen who still wants a baby could just go ahead and have one, the old-fashioned way.

That brings us to the bottom line. Even the most outrageously aggressive effort to prevent human extinction – namely an automated citizen-manufacturing scheme driven by a quasi-religious state propaganda campaign and a tyrannical government, mercilessly wielding the threat of law – even *that* will fail to stop satisfaction sterility. A monumental effort like that would cost a lot of money, and such big expenses have to be justified, sooner or later. Even worse: after a generation or two, the program would create other, whopping social costs as well, because both the citizens and the government would finally come to regard the whole thing as a superfluous nuisance. If such a dire program was serving no useful purpose, except in some abstract sense – in other words if no one actually got anything useful out of it, nor *felt better* knowing that it existed – then it's simply impossible that it would continue to receive public support for long. Some new generation will grow up, look at the program, shake their heads and wonder what on earth was on their grandparents' minds. Then they'll just close the whole thing down.

Here's what that tells us. The problem is not ultimately a practical one ("Who will make children?") but rather a motivational one ("Who will *want* children?"). That sounds a little weird to our ears, because we look around us – or even just look in the mirror – and we see people who *do* want to have children. In fact, the great majority of us feel a distinct horror at the notion of a future in which children and young people have simply ceased to exist. But remember, just six or seven generations ago, a worldly European could feel utter shock at the idea of any married couple exercising any form of family planning. In Leo Tolstoy's 1877 novel, *Anna Karenina,* Anna is talking with her young friend, Dolly. The dialogue can be paraphrased like this:

> **Anna**: I shall have no more children.
> **Dolly**: How can you say that?
> **Anna**: I won't have any because I don't wish to.
> **Dolly** [*with a look of horror*]: Impossible!
>
> > *Dolly thinks about it, and suddenly understands "all those families, hitherto incomprehensible to her, where there were only one or two children."*
>
> **Dolly** [*in French*]: Isn't that immoral?

That may be the first mention of family planning in any book, in any language. Before that, the concept seems to have been effectively unthinkable, or at least unspeakable. But since Tolstoy's time, childbearing has been losing its sacred role in wealthy societies with almost unbelievable speed.

Despite that, I will suggest that the small families found in developed nations today are still a bit larger than the parents in each household typically desire. Because the drive to have large numbers of children is evaporating so very quickly, I'd like to propose that there's a sort of intergenerational flywheel effect. Here's what I mean by that. To some degree, we of the current generation tend to have children because our parents did so. That's not our only motive, but it's one of them. If our parents felt strongly that three children was just right, then we ourselves might have two... even though our hearts tell us that one would be plenty.

The reason I suggest this possibility is that a number of sociological studies have found direct correlations between unhappiness and fertility among adults. In other words, social science surprisingly shows us that people with more kids are less happy than those who have few, or none. That seems to suggest that people often *think* they want children, so they go ahead and have them, only to realize too late that they didn't want them as much as they'd imagined. That begs the question: Why were they so convinced that they wanted kids to begin with, if they didn't *really* want them? My guess is that they were experiencing a sort of intergenerational momentum. All of us absorb values regarding the appropriate size of a family from our parents, when we are little children. But the truth is that we're a different generation, and our parents' values are not quite right for the times we live in.

Voluntarily sterile people are already a substantial minority in developed nations. That sector of society has grown tremendously in recent generations, and is still growing. For the US generation born in the early 1950s, 10% of people went childless, including, of course, those who did so involuntarily. If we look at people a generation later, those born around the mid-seventies, we find that 18% of them went childless for life. Roughly 5% were involuntarily childless in both of those generations, which means that voluntary childlessness more than doubled (from 5% to 13%) in *one generation*. Today, the world is being handed over to the people born in the late 1990s, who are passing through the peak of their childbearing years at the time of this writing. We won't really have the numbers until they are middle-aged, but the early indications certainly suggest that the 'childfree' demographic is going to be much larger in this generation than the generation before. Satisfaction sterility is building up momentum like a snowball rolling down a mountainside.

The feeling that we somehow have to plan ahead for this future catastrophe, and figure out something to do about it, is as absurd as if we learned of some group of people – even a major social movement – in the year 1700 who heard that the people of the 21st century would be engaged in family planning, and determined that *they* must do something in order to discourage *us*. If we heard about their efforts, we'd find them utterly laughable. We don't want the same things now that people wanted back in 1700! And anyway, what possible leverage could those long-dead busybodies think they can have over *us*? Did they think they were going to show up in a time machine and scold us?

Naturally, if our descendents in the year 2300 hear that we 21st century fuddy-duddies were worried about satisfaction sterility, they'll think that was very quaint and old-fashioned of us. If they hear that we actually talked about the possibility of *doing* something about it, they'll split their sides laughing at us. What possible leverage could we have over them? And why would they want to obey the fusty advice of a bunch of crabby, long-dead ancestors? The world will belong to them, and they can make any mistakes with it that they might feel like making... just as we do.

Only a few generations ago, it would have been hard to find a Western European or American or Japanese person who would have believed a fortune-teller who said: "One day soon, most people won't even *try* to produce the biggest family possible." They would have assumed that such an attitude was implausible in our species, regardless of the historical circumstances, because it went 'against nature'. Our ancestors would be very likely to think that the value of having a big houseful of children was self-evident, and so important that pretty much anyone in the world would go ahead and at least try to fill up the house with kids – the more, the better. They'd be confident that people would always do that, even if they couldn't afford to do it the easy way, and even if it meant a terrific strain in terms of household labor and expenses. Most of us today, looking back on those views, find them so outdated and alien that they're kind of comical, in a quaint, nostalgic sort of way.

Today, there are still parts of the world in which mean per capita wealth is as low as it was in the late 18th century United States, and in most such places, very large families are still the norm. So it's not the progressing of time that makes our attitudes so different from those of our ancestors; it's more a matter of getting access to conveniences, luxuries, and other sources of satisfaction. With increasing personal access to satisfaction, whether across historical time or geographical space, we see a loss of the desire of individuals to have a lot of children. We modern citizens of the wealthy nations haven't stopped filling our houses with ten or fifteen children because it's not fashionable, we've stopped doing it because we no longer want to.

Imagine me speaking through a magic telephone with one of my ancestors of ten generations ago. He's surprised to hear that I don't want ten or fifteen kids. I explain that that's commonplace in my era. He's amazed. I explain that the world is more crowded now. He points out that that's irrelevant to my motives, and I admit that's true. I tell him that kids are expensive now. He laughs in my ear. I get tired of arguing, and suggest we agree to disagree. He is outraged: my views are flat-out *wrong*... and against nature! I finally have to hang up on him.

The magic phone rings, and now I find myself talking with my descendent from ten generations in the future. I'm surprised to hear that he has absolutely no interest in having a family, and is perfectly content to just own a cat. He explains that that's commonplace in his era. I'm amazed. He explains that the economy is pretty thoroughly automated in his day, so society doesn't need many workers. I point out that that's irrelevant to his motives, and he admits that's true. He tells me that kids are expensive in his day. I laugh in his ear. He gets tired of arguing, and suggests we agree to disagree. I'm outraged: his views are flat-out *wrong* and against nature... people with views like his are going to cause the human species to go extinct! He gives an exasperated sigh and hangs up on me.

There's nothing we can do to prevent the changing human attitude toward reproduction. That's why our extinction is inevitable.

10

THE EXTINCTION SPIRAL

The previous chapter pointed out that the reason we humans are going to go extinct is not because we lack the means to do anything about it, nor that we lack the will to prevent it from happening. The problem is that, in the future, we are *going* to lack the will to do anything about it. That progressive change in our will is already happening, and has been happening quickly and steadily for a long time now. It has been misidentified by experts as a late phase of demographic transition. But, in fact, it's an independent phenomenon called satisfaction sterility, a natural response of the brain to progressive innovation.

That leaves us in an awkward position, as we try to figure out how satisfaction sterility will develop in the future, as a social trend. We can't just sit in armchairs and wave our hands, giving our personal opinions on the matter – not if we want to have any chance of getting it right. Since the very nature of the trend is that it represents an evolutionary change in human desires and motives, we can't safely make any assumption that future generations are going to feel and act the same way we do. It won't do much good to project ourselves forward in time in our imaginations, and say, "Well, here's what *I* would do..."

The only way we can answer the question, "Could the human desire to have children really dwindle to zero in future generations?" is to step back from the issue and examine it as an abstract question, using impartial reason and the best data we can get our hands on. And if that's going to be our method, then we're in store for some good news. We have at our disposal *huge* collections of statistics showing how our prosperity and birth rates and death rates are changing. These are among the biggest datasets ever collected by humankind, for any purpose. By some estimates, the act of gathering all these data – including records of the births and deaths of billions of people in every corner of the world – required more man-hours than building the pyramids of Egypt. Having all those data at our fingertips makes it relatively easy to model the future of our species – at least, in broad terms. This chapter will present a simple model based on those sets of data, carrying the trends forward in time so we can see what's going to happen to our species.

But first, what sort of 'model' are we talking about, here? A model of this sort consists of a set of equations that mimic what we actually observe happening in human populations of the past and present. We feed various social and economic factors into the equations, and they cough out estimates of the birth rates and death rates seen among people of each sex and age group in the population. If we can put together a set of equations that use realistic assumptions about what's happening in society, and that also spit out results that are a good match to the actual birth rates, death rates and population sizes seen in real life, then we've got a working model. Then all we have to do is plug in today's world population (7.3 billion people), and tell a computer to calculate how many people there are going to be next year. Then we do the next year, and the next, and so on.

All the details of my model are described in Appendix One. In fact, the model itself is available for anyone to see, at an Internet address that's also given in the appendix. Here's a thumbnail description of how the model works, and the assumptions it makes.

The main engine that drives the model is rising prosperity. Very few of us are aware of our rising prosperity, worldwide, unless we really stop to think about it. It's part of human nature to feel that our economic situation is deeply dissatisfactory. If you watch the news, you probably tend to think of our society as wobbling between rare, precarious growth spurts and long, ruinous recessions. Similarly, most of us individually think that our personal income is inadequate for our needs, even if we are making substantially more than the world average, which is $14,400 a year. The reality, in the US in 2015, is that a single person making $30,000 a year is quite rich. But it sure doesn't feel that way!

If we want to make a model that works, we'll have to drop our natural feelings of financial dissatisfaction for a few moments and recognize the peculiar truth. The truth is that we, as a species, have been getting richer at a spectacular rate (almost 2% per year in real terms) for a very, very long time. Since this economic growth is driven by innovation, which is an unlimited resource, there's no reason to imagine it's going to slow down. The model assumes that this growth is going to continue for a few centuries... which is to say, until we go extinct.

Rising wealth has two big effects on human population, in the model. First, it makes death rates go down, causing people to live longer. Second, it causes satisfaction sterility. The effect on death rates is complicated, so the model divides people into three age groups, to get a close fit with what we see in real populations. Children under five are subject to infant mortality, which can be very severe in poor nations. Older children and young adults, in the age range of five to 24, enjoy the lowest annual death rates seen in most populations. From age 25 to 69, mortality rates are a bit steeper, but still they are often quite steady. From about age seventy on, mortality rates show the effects of old age. These old-age effects are idiosyncratic, so the model doesn't attempt to simulate them; instead, the equation for adults, ages 25-69, is simply extended forward to age one hundred, and then everyone dies. This simplification is unrealistic in principle, but turns out to have very little effect on the actual outcome. So, once the three mortality equations are in place (for infants, youths and adults), the model fits very closely with observed death rates in human populations in the real world.

The effect of rising wealth on fertility is easy to observe and measure in real data sets, so it's easy to add satisfaction sterility to the model... in principle. The job is made a bit more complicated by the fact that older women turn out to be affected by satisfaction sterility more strongly than young women. In fact, in the richest countries, women in their forties produce very few children indeed – something that's not true in poor nations. So, to realistically model fertility, we must divide women into five-year age brackets, spanning the childbearing years from ages 15 to 49. Each of these seven age groups of women gets its own equation for satisfaction sterility. Once we've done that, the model fits the real world very closely.

So, there we have a close-fitting model of human population trends, and one that is remarkably simple in its structure. What happens if we tell a computer to turn the crank, so to speak, and click the years forward one by one? The details of that story are given in Appendix One, and the results can be seen graphically there, in Figure A4. But here are the essential results of the model's mechanical advancement of time:

We've already seen that our population is growing at a ferocious rate right now, regardless of satisfaction sterility. That will continue for another two generations, hitting a peak population of 10.1 billion in the year 2069. But three generations from now, the world population is going to stop growing and begin to plummet. By the 13th generation, the world will be down to a tiny fraction of its current population. Our numbers will continue to decline, and the model predicts extinction in the year 2563. If you're the kind of person who takes life one day at a time, then that probably sounds like a long way off in the future... but from a historical viewpoint, it's right around the corner. We humans have been around for six thousand generations, and we've only got 22 left to go. That means that 99.6% of the human story is already behind us.

The model predicts that a grand total of 7.7 billion more babies are going be born in our species's future, spread out over the entire period from 2016 through our extinction. That number is barely larger than the current world population. It's estimated that a grand total of 108 billion people have *ever* been born, since our species first diverged from *Homo heidelbergensis* 130,000 years ago. If we put those numbers together, we see that 93% of all human births have already occurred, and only 7% are left to go. Not only that, but fewer than half a billion babies are ever going to be born, from this day forward, among all the nations currently categorized by the UN as 'High Income' – such as the countries of Western Europe, the US and Japan. The other seven billion births will happen in other nations, though it's worth remembering that all of those nations are going to become rich as time goes by.

By 2400, the world population will be back down to the Paleolithic level, with about five million people living worldwide. That's less than a tenth of one percent of our current population. But the lifestyle of those future people certainly won't be Paleolithic... far from it! The world will be inhabited, however sparsely, by rich people living lives of leisure in a highly automated economy. By then, hardly anyone will have children. Strangely enough, hardly anyone will *want* them. In the year 2513, 22 generations from now, the youngest woman in the world will have her fiftieth birthday, and she will be childless. That will leave the world with no children and no women of childbearing age... so our species will be effectively extinct. The 3000 elderly people who are still alive on that melancholy occasion will live out the rest of their lives, dying one by one, with no hope of contributing genes to future generations. Nor will they feel any desire to do so.

How literally should we take all these numerical predictions of the model? Not too literally, please. I chose to entitle this book *22 Generations* as a dramatic touch, but there's no way that a model – any model – can reliably predict how long our species has left. The main strength of this model is its extreme simplicity, combined with its excellent fit to the data we've already got at hand. It certainly doesn't account for every possibility, nor for every factor at play in the human world. Historically, some oddball events (like the Black Plague) had real consequences on the world's population trajectory. But seen on a long, multi-generational timescale, even events of that magnitude came and went without changing the outcome very much. Plague or no plague, we're here today – all 7.3 billion of us. By ignoring details, even some really *big* details, this model serves as a robust demonstration of an underlying principle. It's not supposed to be a crystal ball; it's heuristic.

What we *can* take literally is the trend toward satisfaction sterility, its unstoppable nature, and its inevitable outcome. Satisfaction sterility really is going to drive us extinct, and there really isn't anything to be done about that. We can also trust this model to give us a general sense of the time scale. I think it's safe to say that our extinction is going to occur in a matter of centuries rather than millennia – and certainly not millions of years.

There are many models of future population, though I believe that mine is the first to show the long-term effect of satisfaction sterility. Currently, the most widely trusted models of the future of human population size are those published periodically by the United Nations. The UN's Low Fertility Estimate gives results similar to my model, despite being calculated differently. In fact, as I describe in Appendix One, the UN thinks there's a 20% chance that population will fall even faster than my model suggests.

So, although 22 generations may not be a precise prediction, it's a pretty good ballpark estimate for the amount of time our species has left. How far away is 22 generations? Our extinction lies about as far in our future as Columbus or Leonardo da Vinci lie in our past. If you cast your mental eye back as far as the time of Marco Polo, and then turn the other way to gaze that far into the future, you find a world where human extinction is already complete, and the last ruins of our civilizations are already buried under spreading fields and forests. Look a little further into the future and you will find us only in that place where countless other world-dominating species have already gone before us: the fossil record.

Here's an odd bit of consolation. It's possible that the events predicted by this model, and this book in general, might not literally be the extinction of *Homo sapiens*, but only of those of us in post-agricultural societies. Worldwide, there are probably still a few thousand people living in hunter-gatherer bands. If any of them are still around in 22 generations (and that's a very big if), the end of civilization will come as welcome news for them. They'll spread slowly as the world reforests itself, and within a hundred generations or so, their population will probably stabilize at a few million, just as it was thousands of generations ago. As long as those future bands never discover agriculture, our species may survive for millions of years to come. But, again, that happy ending depends on there still being unaltered hunter-gatherer bands a few centuries in the future. I wouldn't count on that. Maybe ignorance can be bliss, but if so, it's a very fragile blessing. As soon as people get their first whiff of the luxuries and joys of civilization, they generally want in.

Another subject that hasn't come up, but bears mentioning, is the notion that we're approaching a technological 'singularity'. That idea was first suggested by mathematician John von Neumann way back in the 1950s. He pointed out that if strong artificial intelligence (AI) is ever achieved, it may develop to the point where it is able to innovate more effectively than we humans do. In that case, it might create an improved subsequent generation of AI, which might create an even better one, leading to a runaway cycle of ever-improving machine intelligence. In that situation, we humans would soon find ourselves hopelessly outclassed. In 2014, both the physicist Stephen Hawking and the high-tech billionaire Elon Musk warned that such an event could lead to human extinction.

But there is another, more insidious danger in strong AI, and one that is much more likely to be a real component in our upcoming extinction. The problem lies in the fact that there is nothing in the nature of our human *intelligence* that makes us want to reproduce. Rather, we reproduce because of biological drives – motivational circuitry that originated much earlier in our evolutionary history than did our intelligence. Half a billion years ago, all my ancestors were fish, and back then fish were just as stupid as they are today. But their reproductive drives were at least as strongly programmed as mine... possibly stronger. In our human evolutionary history, intelligence was tacked onto our genetic natures very late in the game. The huge human brain was an expensive and experimental gimmick added onto a set of well-honed survival mechanisms that had been working just fine without it.

It's not necessarily the case that intelligence must work *against* the project of reproductive success, but it would be absurd to claim that the reproductive drive is somehow a component of intelligence. It's easy to find concrete examples to demonstrate that. Plenty of people have very strong intellectual drives but no effective reproductive drive, and that includes some people who are major innovative contributors to modern civilization. Of course, many of those people have strong sex drives, and many also have strong care-giving drives. They're just not inclined to have children.

In Chapter Eight, we explored the role of automation in preventing future birth-dearth recessions that might otherwise halt the progression of satisfaction sterility. There's nothing new about the idea that machines might take over human jobs: automation has been replacing human labor for a long time. Back in the Industrial Revolution, machines replaced a lot of human brawn in the workplace. More recently, automation has been replacing a lot of menial mental labor, due to the spread of workplace computers. What if a third phase of automation comes along, and artificial intelligence takes over the creative labor of innovating? That would be the technological singularity.

Most commentators who have expressed worries about such an event, such as Hawking and Musk, have been concerned that AI systems will overpower us once they've become smarter than we are. That's not a trivial worry. Still, my opinion is that we can probably rest easy on that account. By the time human labor is completely supplanted by machines in all sectors of the economy – even the labor of innovating – our human interest in generational replacement will be tiny, and falling fast. Economies will be booming, labor will be all but obsolete, and everyone will be too busy enjoying life to bother having children. The smart machines won't have to steal the world from us. We'll abdicate.

So, when the last humans walk the earth, just a few hundred years from now, what will life be like in their communities? What would it be like for me to meet one of the last men face to face, to grab him by the shoulders and shake him, try to make him see sense? I suppose he would just look at me, and blink. His world is just fine, and he can't for the life of him understand why I have such an issue with the thinning of global population in his future time. Anyone who is raising a child in *his* day probably isn't quite right in the head. Life as he knows it is so full of satisfying pleasures and luxuries that anyone who found himself or herself saddled with the care of a child would miss out on an awful lot. In that future world, you can't hire someone else to do it, either... *everyone's* too busy living the good life. And other than the vague specter of extinction at some even more distant future date, there's no really pressing downside to the scarcity of the younger generation. The world's economies are so automated in this future world that the labor pool is effectively redundant.

One of the greatest ironies of this analysis of human civilization and its impending doom is that we finally end up just shrugging off all those pessimists who have foretold grim futures: the social dystopians of the left, the social dystopians of the right, the ecological alarmists. It turns out that our end is going to come about as a direct result of our achievement of unlimited success. We'll just keep climbing and striving, onward and upward to the very peak of human happiness. As we go, our population will gradually thin out through peaceful attrition, each of us dying comfortably, one by one, in prosperous old age, until the last person is left sitting alone at the absolute pinnacle of human comfort and satisfaction. And then there shall be none.

Our species is going to die happy. Indeed, our happiness and its resultant willful infertility will be the cause of our extinction. So I suppose that the story I've told you in this book can't exactly be called a tragedy – except in the strict sense that all the characters are dead at the end of the final act. Even so, strange to say, the story has a happy ending.

AFTERWORD

It took Charles Darwin decades to work up the guts to publish his theory of evolution by natural selection. He knew, as surely as night follows day, that an awful lot of people would misinterpret his motives. He had in his head a scientific observation that seemed obviously true, and he wanted to share it with the world. But he was afraid that various persons would accuse him of pursuing some kind of hidden agenda. He was right, of course: Even today, a lot of people think Darwin was driven by a hidden agenda, perhaps one aimed against traditional religious faiths, and it makes them furious.

I, too, have an uncomfortable feeling that my book is going to make some people leap to the conclusion that I have a secret agenda of some kind. So, I want to go on record as saying that anyone who believes that this book presents some sort of call to action should read it again, more carefully this time. You'll find that I've taken pains to repeat several times that I don't think anyone should do anything to correct the trend toward satisfaction sterility. I think, in time, that people *are* going do a lot of things to try to enhance fertility rates, and I want to be the first person on record to say that they are fools. They're going to make a lot of people unhappy, and they will do no long-term good at all. In Iran, as described in Chapter Nine, that idiocy has already begun.

The problem has two parts: first, none of us want to see our species go extinct, and second, it's in our nature to believe that we can solve any problem we really set our minds to. For the most part, we're right to believe that, but I hope that the arguments of Chapter Nine explain clearly why this case is an exception. And yet, even if we acknowledge that those arguments are convincing, I'll bet most of us nonetheless find the conclusion hard to accept. I think that's because we're not accustomed to thinking of ourselves as the products of evolution, nor of thinking of ourselves as strictly bound by the material world in ways that go beyond the physical limitations of our immediate situation, whatever it might be. We know that we have free will, because every experience reminds us that we can make choices. True, philosophers like to raise nagging doubts about the notion of free will, pointing out that although we can choose what we do, and can even choose to do something that we *think* we don't want to do, we ultimately are limited by our dispositions. We didn't individually create our own dispositions; they were handed to us. The basic human tendencies at the core of our motivational systems, which are shared by almost everyone, are genetic, and all the rest, the details of what we prefer as an individual, are the products of our upbringing and life experiences. So our free will is severely limited, at best. We can make choices, but we cannot select our drives... and those drives are lurking in the shadows *somewhere* behind every choice we make.

That's a hard fact to accept. None of us, probably, is capable of walking through a normal day while keeping that fact clearly in mind at all times. All of us have a psychological need to experience the illusion of personal freedom, and I think that we can only dispel that illusion for short periods. But we *can* do so, if we're so inclined, by thinking the matter through carefully. In that state, keenly aware of the limited nature of our 'freedom', it's not so surprising to find that hardwired, neurological drives exist inside us and rob us of our freedom by giving us a disposition to want some things rather than others. More than that, it's not even surprising to realize that those drives can undergo evolutionary changes as they interact with our ever-changing modern environment.

Furthermore, it's not surprising to realize that by creating the amazing synthetic environments that give civilized people the entire context of modern life, we have created a unique evolutionary experiment – one that is affecting our behaviors in unplanned and unexpected ways. That's what's going to cause our extinction. The only way to prevent our extinction would be to stop the experiment. But since the experiment is civilization itself, that's simply not going to happen.

Why, then, did I write this book, if I don't have any advice to help my species escape its looming fate? My reason for reporting the findings in this book is the same as if another scientist reported that a distant galaxy is exploding – namely, the best evidence indicates that the statement is true. There's nothing to be done about an exploding galaxy, or about human extinction, but there is nonetheless value in knowing the truth. It gives us some added perspective, if nothing else.

So, if I have a hidden agenda, a secret reason for writing this book, that's it: a desire to expand our perspective. Consider this, for comparison: For many children, the realization of their own mortality – the understanding that some day they are going to face personal death – is a crucial moment in growing up. Before receiving that understanding, kids are certainly more carefree, and one might also say more innocent, and perhaps in a superficial sense they're happier, too. But they're also incapable of adult responsibility and adult comprehension of life and the world. It is my sincere hope that the recognition that, as a species, our time in this world is limited will lead us all, as a species, to a higher state of maturity.

Appendix One

A MODEL OF ECONOMICALLY DRIVEN POPULATION CHANGE

As of this writing, this entire model is available for inspection or copyright-free download (in the form of a macro-free Excel file) at www.22generations.com.

This deterministic model uses functions derived from data taken from 20 sampling units. The sampling units consist of two blocks of nations, during the ten five-year periods spanning 1950-2000. One block of nations consists of most of the UN-defined 'High Income Nations'; the other block (here called 'Low Income') consists of most of the other nations. Nations were excluded if complete data were lacking. The included nations, which account for over 91% of the current world population, are as follows:

High Income nations: Argentina, Australia, Austria, Belgium, Canada, Chile, Equatorial Guinea, Finland, France, Germany, Greece, Hungary, Ireland, Israel, Italy, Japan, Netherlands, New Zealand, Norway, Poland, Portugal, Puerto Rico, Singapore, South Korea, Spain, Sweden, Switzerland, United Kingdom, United States, Uruguay, and Venezuela.

Low Income nations: Afghanistan, Albania, Algeria, Angola, Bangladesh, Benin, Bolivia, Botswana, Brazil, Bulgaria, Burkina Faso, Burundi, Cambodia, Cameroon, Central African Republic, Chad, China, Colombia, Comoro Islands, Congo, Costa Rica, Côte d'Ivoire, Cuba, Denmark, Dominican Republic, Democratic Republic of Congo, Ecuador, Egypt, El Salvador, Gabon, Gambia, Guatemala, Guinea, Guinea Bissau, Haiti, Honduras, India, Indonesia, Iran, Jamaica, Jordan, Kenya, Laos, Lebanon, Lesotho, Liberia, Madagascar, Malawi, Malaysia, Mauritania, Mauritius, Mexico, Mongolia, Morocco, Mozambique, Myanmar, Namibia, Nepal, Nicaragua, Niger, Nigeria, North Korea, Pakistan, Panama, Paraguay, Peru, Philippines, Romania, Rwanda, São Tomé and Principe, Senegal, Seychelles, Sierra Leone, Somalia, South Africa, Sri Lanka, Swaziland, Syria, Tanzania, Thailand, Togo, Trinidad and Tobago, Tunisia, Turkey, Uganda, Vietnam, Yemen, Zambia, and Zimbabwe.

There are two major lacunae in these lists. 1) Most of the nations of the ex-Eastern Bloc, especially the Russian Federation. GDP estimates of these nations during the communist years are not reliably comparable with those of other nations. 2) Most of the major oil-producing nations of the Middle East. My main source of GDP data (the OECD; see below) disagrees with other major economic analyses on the calculation of GDP in this region.

The data collected from each sampling unit consist of estimates, largely based on census data of varying reliability. Per capita GDP (gross domestic product) estimates are from the Organization for Economic Co-operation and Development (Angus Maddison, 2003, *The World Economy: Historical Statistics*, OECD), and are expressed in terms of purchasing power parity (PPP). This means that the units, though expressed in 2015 US$, have been adjusted by economists in an effort to show the real spending power of local currency in the country of origin. Other than GDP, all sampling unit data are from the UN Population Division's *World Population Prospects: The 2015 Revision.*

Eleven estimated data were gathered from each of the 20 sampling units:

> Per capita GDP
> Female mortality rate, ages 0-4
> Female mortality rate, ages 5-24
> Female mortality rate, ages 25-69
> Fertility in each of the seven 5-year age groups of females, aged 15-49

In each case, data were averaged per capita over nations and annually over the five years within the sampling unit. This yielded 220 base data, from which I derived Functions 1-10 (see below) by regression. These functions determine fertility and mortality in the model.

The model begins in the year 2015. The initial population values are from female age structure pyramids for each of the two nation blocks in that year, again from the 2015 UN source. The five-year bins were split evenly into one-year initial bins for the model. The initial per capita GDP values for the High Income and Low Income nations in 2015 are from the World Bank's *Databank.*

Derivation of the rate of growth of global per capita GDP

The data in Figure A1 show that the mean value of real annual global GDP growth was 1.98% during the 38-year period 1970-2008, as estimated by the IMF, adjusted by global population. This fits well with longer-term estimates from the OECD.

This model therefore assumes 1.98% inflation-adjusted annual per capita GDP growth for all nations in the future.

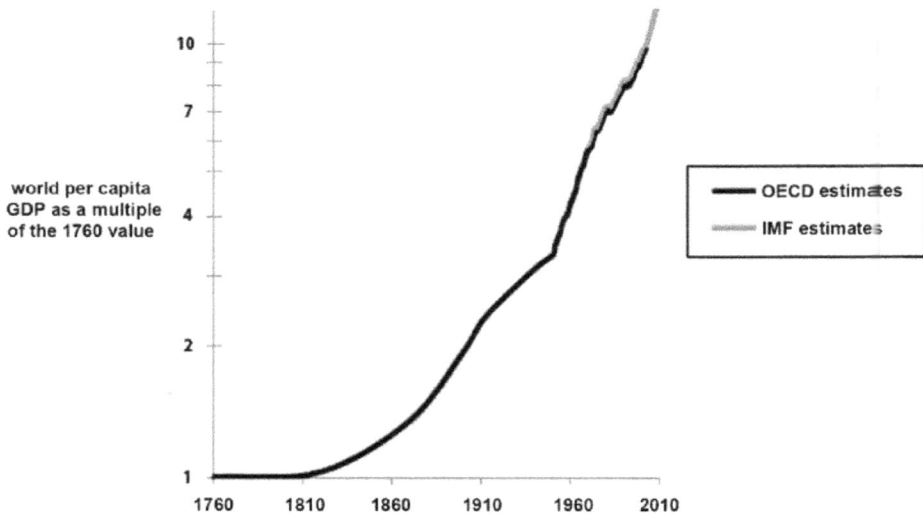

Figure A1. Rising prosperity.

The average person, worldwide, is over ten times richer than at the start of the Industrial Revolution. Money values in this graph are inflation-adjusted. Note the log scale.

Sources: Global GDP and population estimates 1760-2003 from Angus Maddison, 2003, *The World Economy: Historical Statistics*, OECD. Global GDP growth estimates 1970-2008 are from the International Monetary Fund, January 2009, *World Economic Outlook Update*. World population estimates 2004-2008 are from the UN Population Division, *2015 Population Prospects*.

How the model works

The model makes the simplifying assumption of 50% males, 50% females. Each generation is calculated for females only (using halved fertility rates), and then the resulting population is doubled to include males. In real populations, female mortality is usually lower, and this skews the model's annual population projections slightly upward.

The model treats fertility of females in each of the seven five-year age brackets from 15 to 49 as a separate function of per capita GDP (Functions 1-7, described below).

The model treats mortality as having three separate annual, density-independent rates, applying to ages 0-4, ages 5-24, and ages 25-99. Each of these is treated as a separate function of per capita GDP (Functions 8-10, described below). All survivors to age 100 die abruptly. Observed annual mortality rates are usually much higher among infants than adults, and higher in older adults than in youths, so this three-part mortality model fits empirical data quite well in most times and places. However, in the real world, mortality among the elderly follows very different patterns from those seen among non-elderly adults, and no effort was made to account for that difference in this model. The data used for determining the third mortality rate (ages 25-99) were actually those observed among ages 25-69 in real populations, and the model merely extends that rate forward to age 99, without any adjustment factor. This creates a false boost in total population output from the model, but one that doesn't affect the model's internal calculations, because the 'extra' population consists of elderly women who are beyond their childbearing years.

The model stores 200 population bins: ages 0 to 99 for the Low Income nations, and ages 0 to 99 for the High Income nations. These are initially stocked with the observed 2015 values. For each subsequent year, the following steps are taken:

a) Advance the per capita GDP of the High Income nation block by 1.98%, and do the same for the Low Income nation block.

b) Calculate births, using Functions 1-7, and replace the "Age 0" bins with half of the new births (the female babies).

c) Replace each age bin from 1 to 99 with the previous age bin. The "Age 99" bin turns 100 and is discarded.

d) Calculate mortality and remove it from each age bin, using Functions 8-10.

e) Estimate the year's population, by summing the 200 bins, and doubling (to include males).

Derivation of Functions 1-7: Fertility

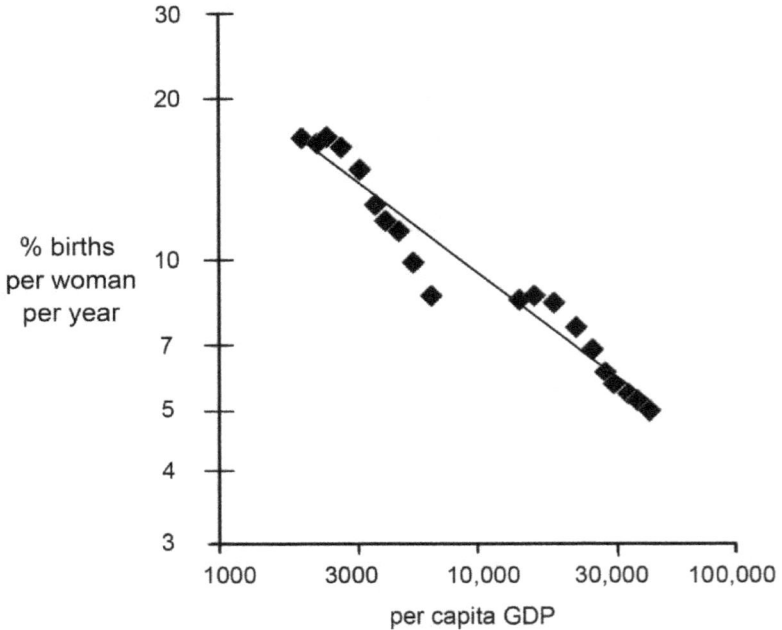

Figure A2. Satisfaction sterility.

As the average woman (age 15-49) in a modern nation gets richer, she also has fewer children. Note the log scales. See text for data sources.

Age-specific fertility is shown here as a percentage: births per year per 100 women in a given range of childbearing ages. The data in Figure A2 show the average fertility among all women of childbearing age (15-49) from the 20 sampling units plotted against per capita GDP on a log-log graph (which allows a power function to be seen as a straight line). The regression closely fits the equation:

$$\text{fertility} = 291 \, (\text{per capita GDP})^{-0.368}$$

The coefficient of determination, R^2, is 95.0%.

To allow for die-off effects in the model among those sampling units in which substantial mortality occurs to women during their childbearing years, best fit regressions to power functions were calculated separately for each 5-year bin of female ages, giving seven age-specific fertility functions, called Functions 1-7:

1) age 15-19 fertility = 171 per capita GDP $^{-0.348}$
2) age 20-24 fertility = 308 per capita GDP $^{-0.314}$
3) age 25-29 fertility = 176 per capita GDP $^{-0.245}$
4) age 30-34 fertility = 284 per capita GDP $^{-0.341}$
5) age 35-39 fertility = 972 per capita GDP $^{-0.541}$
6) age 40-44 fertility = 3800 per capita GDP $^{-0.796}$
7) age 45-49 fertility = 8988 per capita GDP $^{-1.069}$

A reasonable objection to this use of data might be that, during part of the study period, China was implementing an immense fertility-reduction policy. This issue is dealt with in Appendix Two, Topic G, and turns out to be minor. Also, note that although prosperity level accounts for 95% of fertility, the residual 5% also shows an interesting pattern. That residual scatter is explored in Appendix Two, Topic H.

Derivation of Functions 8-10: Mortality

The data in Figure A3 are the 20 sampling units, showing all female deaths in the age ranges 0-4, 5-24, and 25-69, on a log-log graph. The relationship between annual mortality and per capita GDP in each age range is modeled by the closest fit of a power function to the data, giving Functions 8-10:

8) age 0-4 mortality = 12525 per capita GDP $^{-0.804}$
9) age 5-24 mortality = 255 per capita GDP $^{-0.83}$
10) age 25-99 mortality = 23.9 per capita GDP $^{-0.391}$

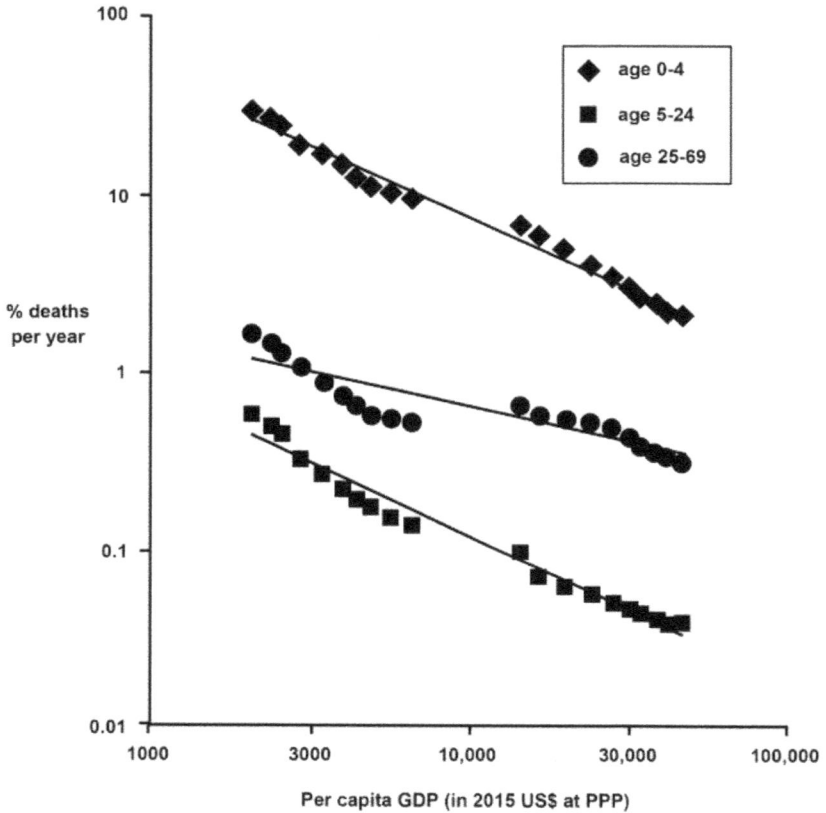

Figure A3. Money and survival.

As an average civilized society gets richer, mortality drops. Note the log scales. Also note that mortality rate is lowest for young people who have survived infancy, and greater for those who are either older than 24 or younger than 5. See text for data sources.

Note, as described above, that Function 10 is calculated to fit real data up to age 69, but is used in the model to generate estimates of mortality up to age 99.

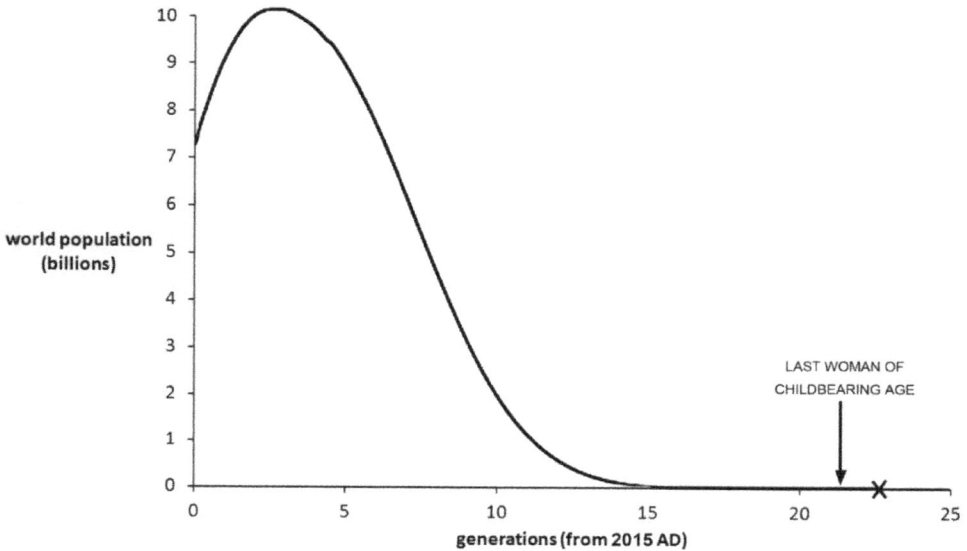

Figure A4. Human extinction.

The simple parameters of this model lead to human extinction after a fairly small number of generations.

Results

When this deterministic model is run according to the parameters described above, it yields the global population projection shown in Figure A4. The model starts in the year 2015 with a population of 7.35 billion. The last child is born 20 generations from now, in 2463. When she turns 50, in the year 2513, without having reproduced, our species is biologically extinct. There are still 3000 people alive at that point, but all are elderly – there are no children, and no women of childbearing age. The last of those old-timers dies in the year 2563.

I do not propose that anyone take this outcome as a prediction of future events on the basis of the model's own strengths. It isn't intended to be 'realistic'; rather, it is a heuristic tool. Its simplicity is its greatest asset: like a simple piece of gearwork, the workings of the model can be easily understood, and so the underlying principle is made clear. The main text of this book is an effort to place that principle in contextual perspective. I *do* propose that our species will go extinct in twenty or thirty generations, but I do so on the basis of this book's arguments, not on the basis of this model *per se*.

Note that the United Nations (using a very different model from mine) gives a 20% chance of a future population trajectory as low as the one in Figure A4, or even lower.

Appendix Two

A CLOSER LOOK AT VARIOUS TOPICS

At the time of this book's publication, all of the calculations used in Topics B, E, G and H are posted as Excel files at www.22generations.com.

Topic A: Very long-term economic growth.

Figure A5 is provided to bolster the assertion, both in Appendix One and in the main text, that per capita prosperity has been rising worldwide for a very long time.

As mentioned in Chapter Six, prior to about 1700, no society on earth had ever been so prosperous that the average individual was free of the most straightforward constraint on reproduction: resource availability. (This is also explored in more detail below, in Topic B). Prior to 1700, in every nation, the average citizen was making less than the equivalent of about $5000 a year in 2015 US dollars; at that income level, he or she could have more children – or at least more who survived to adulthood – if he or she could make more money.

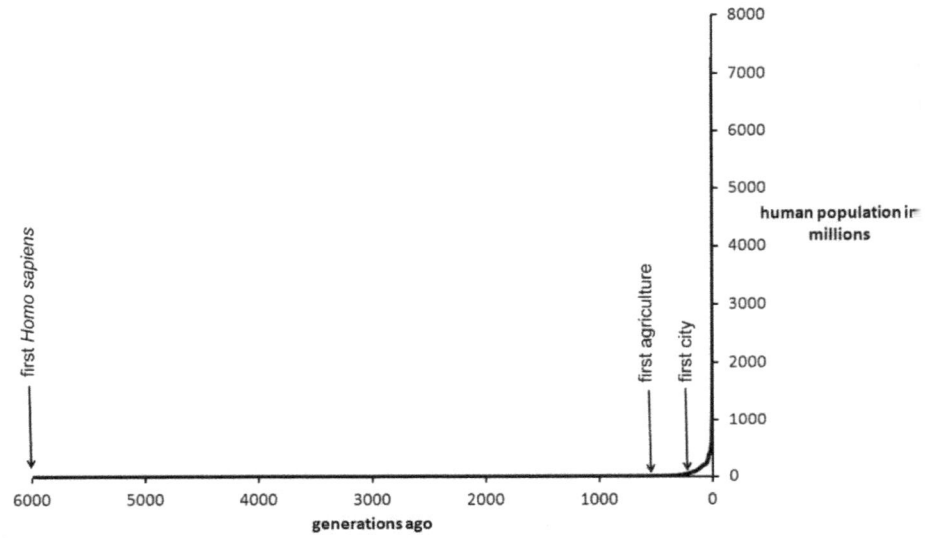

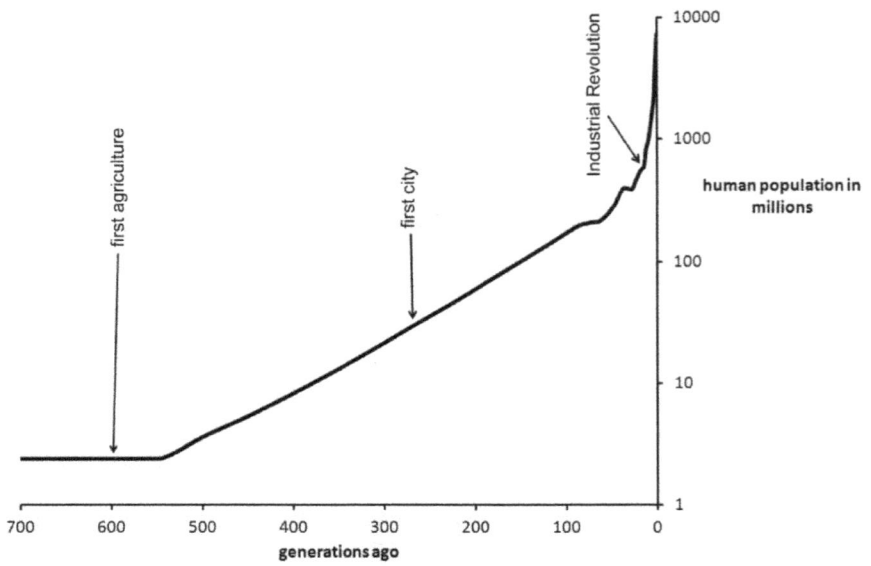

Figure A5. Human history.

These two graphs show human population (in millions) over the course of past generations (where one "generation" equals 22 years). Key events are indicated: a) the first appearance of our species from a variety of *Homo heidelbergensis*, 130,000 years ago; b) the first agriculture (of grains in the Middle East), 10,000 years ago; c) the urbanization of the first true city, Uruk of Mesopotamia, 6000 years ago; d) the beginning of the Industrial Revolution around 1760.

The upper graph uses a linear scale, so the exponential population explosion since the innovation of agriculture hugs the axes closely. The lower graph uses a log scale and gives a close-up of the past 700 generations. Human population size has not stabilized since agriculture appeared.

Sources: Data 1950-2015 from US Census Bureau; data and estimates before 1950 from K. Klein Goldewijk and G. Van Drecht, 2006, *HYDE 3: Current and historical population and land cover*, pp. 93-112 in Eds. A. F. Bouwman, T. Kram, and K. Klein Goldewijk, *Integrated modeling of global environmental change. An overview of IMAGE 2.4*, Netherlands Environmental Assessment Agency (MNP), Bilthoven, The Netherlands.

Since that (very Darwinian) situation applied through most of human history and prehistory, the graphs in Figure A5, showing the rapid growth of human populations in the past 600 generations, can be interpreted as indicating that prosperity has been on the rise throughout most of that period. No economist is likely to go out on such a thin limb as to give actual dollar estimates of prosperity for people living (say) 5000 years ago, but we *can* make tentative estimates of world population back then. The fact that it was rising is a very strong indicator that civilized societies were providing their citizens with increasing prosperity.

The firmest estimates of human population in the past come from the most recent 2000 years (thanks to archaeology) and from the Paleolithic and Neolithic periods prior to 12,000 years ago (thanks to paleontology and physical anthropology). The period between 600 and 100 generations ago is vaguer, and the modelers who provided the data shown in Figure A5 use a simple exponential growth curve to fill the gap (the straight line seen in the middle portion of the lower graph). Fortunately, the arguments found in this book don't rely much on the exact shape of that portion of the curve.

Topic B: The post-Darwinian threshold.

In Chapter Six, I pointed out that once a society's prosperity rises above a certain level, a large majority of female babies can be expected to survive through most of their childbearing years (that is to say, to age 40 or so). Beyond that level of wealth, the average family will not experience much added Darwinian fitness by increasing their prosperity even further. In other words, further increase of wealth may make them more comfortable, but it will not substantially increase the representation of their DNA in future generations. I refer to this transition level of wealth as the post-Darwinian threshold.

Figure A6 presents data showing that, although the transition is continuous, the curve does in fact pass through a fairly well-defined shoulder. In societies that are poor enough that fewer than 85% of newborn girls can be expected to live to age forty, there is a very steep regression between

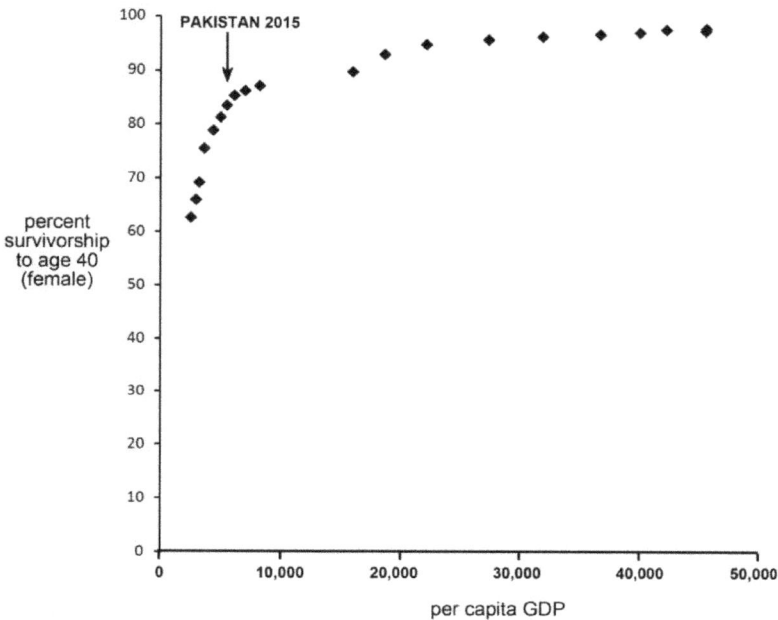

Figure A6. Beyond Darwin's reach.

The 20 data sampling units described in Appendix One are shown here (consisting of the High Income and Low Income nations during the ten 5-year periods from 1950 to 2000). In nations poorer than modern Pakistan, a typical woman's chances of surviving from birth through the bulk of her childbearing years (to age 40) is very strongly dependent upon per capita wealth in her society – just as Darwin would have predicted. But any nation richer than that can only increase the survivorship of childbearing females very slightly, because most of them will survive anyway. When a nation is as rich as modern Pakistan, Chile or Hungary, over 85% of women survive through their main childbearing years. Beyond that level of wealth, prosperity has almost nothing to do with natural selection.

Sources: Estimates of female cohort survivorship at age 40 from UN Population Division, *World Population Prospects*, 2015 Revision. Per capita GDP (expressed in 2015 US$ at PPP) from Angus Maddison, 2003, *The World Economy: Historical Statistics*, OECD.

per capita GDP and survivorship through the main childbearing years. In other words, in these very poor societies, families with a little more money have an advantage over other families, in strictly Darwinian terms: they are likely to have more descendents in future generations. But the regression becomes very shallow as wealth increases to the right of the curve's shoulder. In that region of the graph, increased prosperity has only a marginal affect on a family's headcount of grandchildren, great-grandchildren and so on.

This shoulder in the curve marks the post-Darwinian threshold. Societies that are richer than the threshold amount are largely cut off from natural selection, as discussed in Chapter Six. What may surprise some readers is that the threshold occurs at a per capita GDP of about $5000, in 2015 US dollars. That's well below the official poverty line in the US, as calculated in the following way:

Average household size in the US is 2.58, according the 2010 US Census. The US Department of Health and Human Services defined the 2014 poverty line as an annual income of $19,790 for a household of three and $15,730 for a household of two, which works out to $18,084 for a household of 2.58. Each of the 2.58 household members accounts for $7009 a year, putting them far beyond the post-Darwinian threshold of $5000. So they may be poor, but they are still extremely well protected from natural selection.

Topic C: The spread of a hypothetical fecundity gene in an urban population.

gen	Pop., million	FF-FF couples (1000s)	Ff-Ff couples (1000s)	ff-ff couples (1000s)	FF-Ff couples (1000s)	FF-ff couples (1000s)	Ff-ff couples (1000s)	percent houses with F
0	1	0	0.048	490.3	0	0.012	4.8	0.99
1	0.78	0	0.38	367.9	0.002	0.058	11.9	3.24
2	0.67	0.001	2.8	277.6	0.044	0.44	28.0	10.15
3	0.73	0.04	17.3	217.6	0.8	2.8	61.4	27.45
4	1.15	0.99	78.8	199.1	8.8	14.0	125.2	53.36
5	2.58	12.2	281.4	244.1	58.6	54.6	262.1	73.26
6	7.06	88.7	896.4	401.5	281.9	188.7	599.9	83.66
7	21.16	486.6	2745.2	806.7	1155.7	626.5	1488.1	88.96

The table above shows a simple model of the spread of a hypothetical fecundity gene in an isolated human population, as described early in Chapter Eight. We imagine a city that initially has one million people, and no immigration or emigration. In each generation, everybody finds a spouse at random and has children. If neither parent carries the dominant 'fecund' allele (F) for the fecundity gene, then the household produces 1.5 children, but if either parent does carry the F allele then the household produces 10 children. Initially, 0.49% of all alleles for this gene are F, so that just under 1% of households initially have a copy of the allele in at least one of the two parents. Allele frequencies are calculated by probabilistic Punnett squares, using the Hardy-Weinberg equation:

$$\text{freq}(F)^2 + 2[\text{freq}(F)][\text{freq}(f)] + \text{freq}(f)^2 = 1.$$

Some notable results are seen in the two heavily outlined cells of the table. After four generations, over half of the households are carrying the F allele. In the seventh generation, the city's population is over 20 million, and exploding. As described in the main text, these results dramatize two important points about fecundity genes in modern human populations:

1) They are extremely unlikely to exist, because their existence would be so obvious, and yet they have not been discovered. 2) In the unlikely event that they do exist, they are likely to drive us quickly to a far uglier extinction (due to crowding) than the comfortable extinction we are currently facing due to satisfaction sterility.

Topic D: Demographic transition.

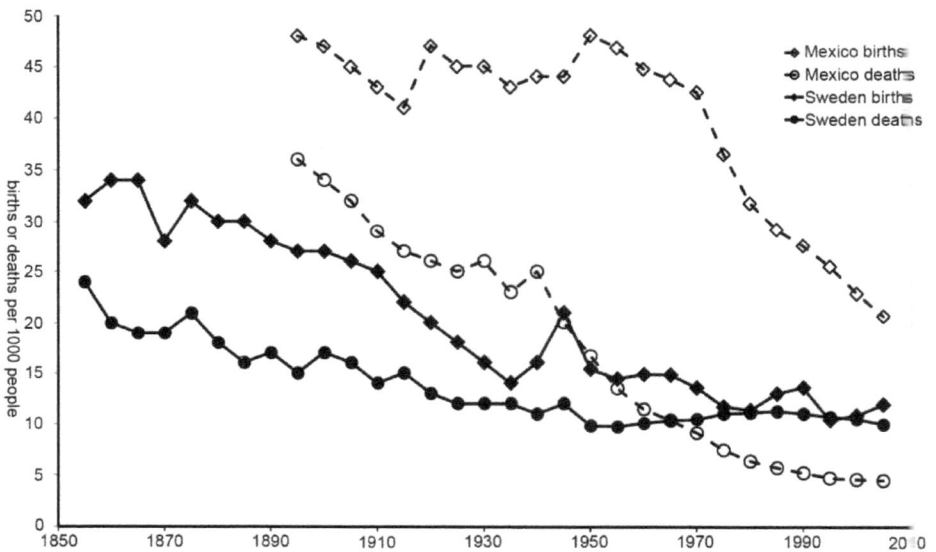

Figure A7. Salvation by demographics?

Typical data showing the fall of death rates and birth rates in a more developed nation (Sweden) and a less developed nation (Mexico) as prosperity rises. Although Mexico's "transition" came 80 or 90 years later than Sweden's, both show remarkable similarities. First the death rate falls and causes population growth, and then, two or three generations later, the birth rate falls and puts an end to the population explosion.

Sources: UN Population Division, *World Population Prospects: The 2015 Revision*; and US Census Bureau, *International Database.*

Figure A7 shows two typical examples of what is commonly known as 'demographic transition', one in a country that is currently classed as High Income by the UN (Sweden) and one that is still developing (Mexico). In both cases, as prosperity rose beyond a certain point, there was a strong drop in the death rate, and then (about two or three generations later) a fall in the birth rate. Mexico's death rate fell during the early 20th century, and its birth rate is still falling now. Sweden's death rate fell in the early 19th century, followed by declining birth rates in the late 19th century.

Although the three phenomena of rising wealth, falling death rate and falling birth rate clearly have some important links, there is no evidence that all three are part of a single process rather than two separate processes. The discussion in Chapter Eight shows that there is no direct link between falling death rates and falling birth rates, and that each is independently caused by rising prosperity. If that's correct, then the term 'demographic transition' is deeply misleading and should be avoided. Falling death rate is a major facet of socioeconomic development. Falling birth rate is a symptom of satisfaction sterility.

Topic E: The relationship between per capita GDP and total fertility rate among modern nations.

In Chapter Nine, I mentioned that the first fertility-enhancement laws are being mooted at the time of this writing, in Iran. This news might raise some eyebrows, because Iran is considered a Less Developed nation by the UN, and many readers may still hold the outdated belief that such nations are hotbeds of overpopulation. Is Iran some sort of freakish statistical outlier? Let's see.

Figure A8 shows the power regression between per capita GDP and TFR, or total fertility rate, among nations that are above the post-Darwinian threshold. In other words, these are the nations with per capita GDP of $5000 or higher, in 2015 US$ at PPP. China has been removed from consideration because of their artificial manipulation of TFR (through their one-child policy), and the nations of the ex-Eastern Bloc have been removed because of regional issues discussed in Topic F, below. The strength of the regression between TFR and per capita GDP among nations in Figure A8 is modest ($R^2 = 0.572$), but the trend is evident and the slope is strong.

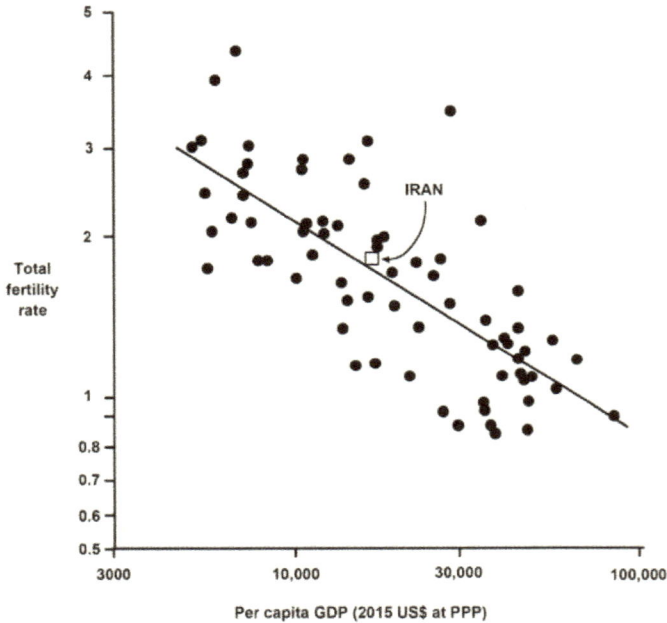

Figure A8. Iran's fertility crisis.

Although it might seem odd that a UN-defined Less Developed nation like Iran should be experiencing satisfaction sterility, it's actually right on target. Iran's total fertility rate (TFR) is almost exactly at the value predicted by the nation's level of prosperity. Note the log scales.

Sources: UN Population Division, *World Population Prospects: The 2015 Revision*; World Bank *Databank*.

Iran's position can be seen to fall very near the middle of the scatter, almost directly upon the trend line. Iran is certainly not a statistical abnormality; Iran's collapsing population is typical of any nation at that level of wealth, and richer nations typically show the syndrome even more strongly. Roughly speaking, any nation which achieves a per capita GDP of $10,000 should expect to lose the ability to regenerate its own population, and begin the slide toward annihilation. Iran, with a per capita GDP of $16,500, is well beyond this point of instability. For that matter, the world as a whole, with a per capita GDP of $14,400, is far beyond the point of no return.

Topic F: Religion as an enhancer of fertility.

As discussed in Chapter Nine, various religions have managed to enhance birth rates among their adherents, both through propaganda and extralegal command structures. Catholicism and Islam are cited as examples in the text.

Figure A9 shows some data suggesting the powerful influence of Muslim religion as an enhancer of fertility in the nations that were once components of the Eastern Bloc. There are now 21 such nations spreading across Eurasia from the Mediterranean to the Pacific. Six are predominately Muslim by population, and the other fifteen are predominately non-Muslim. As seen in the figure, the six Muslim nations fall in a conspicuously separate cluster from the others, due to higher birth rates. Furthermore, each cluster of nations, taken separately, shows a clear, simple pattern of satisfaction sterility: a negative power relationship between per capita GDP and birth rate. Within each cluster (Muslim and non-Muslim), increasing prosperity causes decreasing fertility. But at any given level of wealth, birth rates in the Muslim nations are roughly double what they are in the non-Muslim nations of the region.

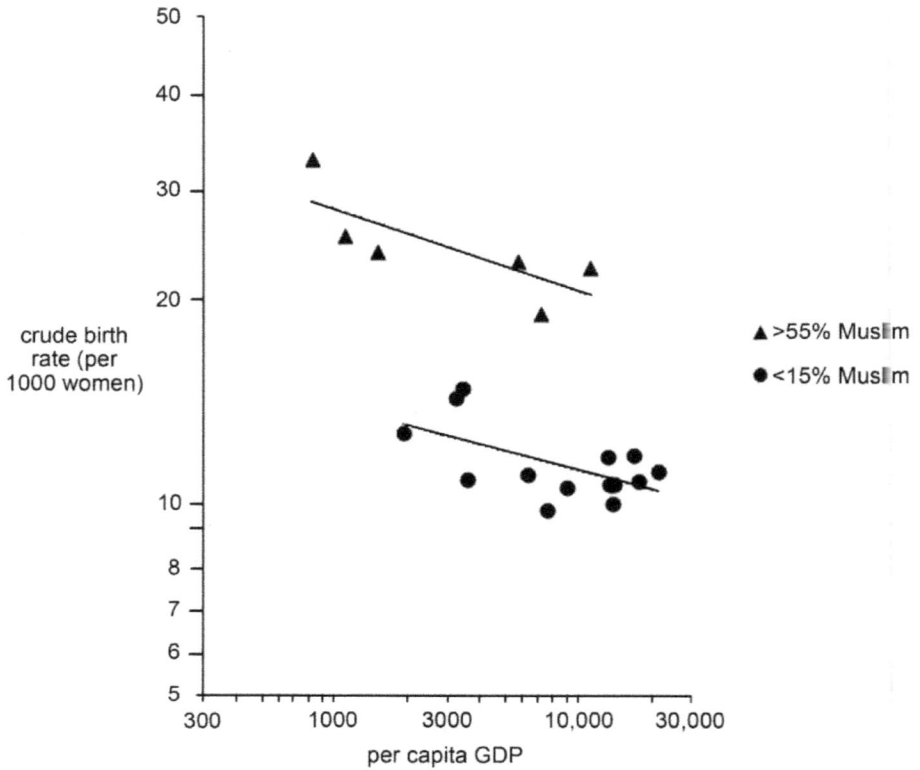

Figure A9: Muslim influence on fertility.

The 21 nations of Eastern Europe and the ex-Soviet Republics can be divided into six that are >55% Muslim by population, and fifteen that are <15% Muslim. There's nothing in between. Both groups of nations show satisfaction sterility, but the predominately Muslim nations have much higher fertility at all levels of prosperity. Note the log scales.

Sources: Birth rate data (2005-2010) from the UN Population Division, *World Population Prospects: The 2015 Revision*; Muslim census data from Pew Research Center, 2011, *The future of the Global Muslim Population*; 2011 per capita GDP data (in 2015 US$) from the World Bank *Databank*.

Topic G: The effect of China's one-child policy on the data used in the model.

In Appendix One, the relationship between per capita GDP and fertility (Figure A2) was presented with the caveat that China's one-child program might have substantially skewed the data. Figure A10 examines that possibility.

China's one-child policy has been implemented since 1980. Estimates vary widely as to the success of the policy, in terms of prevention of births. Probably the highest estimates are those of the Chinese government, claiming that up to 300 million births had been prevented by the year 2005 (see Therese Hesketh *et al.*, 2005, *The Effect of China's One-Child Family Policy after 25 Years,* New England Journal of Medicine, 353: 1171-1176.) That would amount to 12 million per year.

As seen in the figure, giving a boost of 60 million births to each of the four 5-year Low Income nation samples between 1980 and 2000 does not strongly change the power regression between per capita GDP and fertility. However, the shift greatly improves the fit of the regression, raising the R^2 from 95.0% to 97.6%.

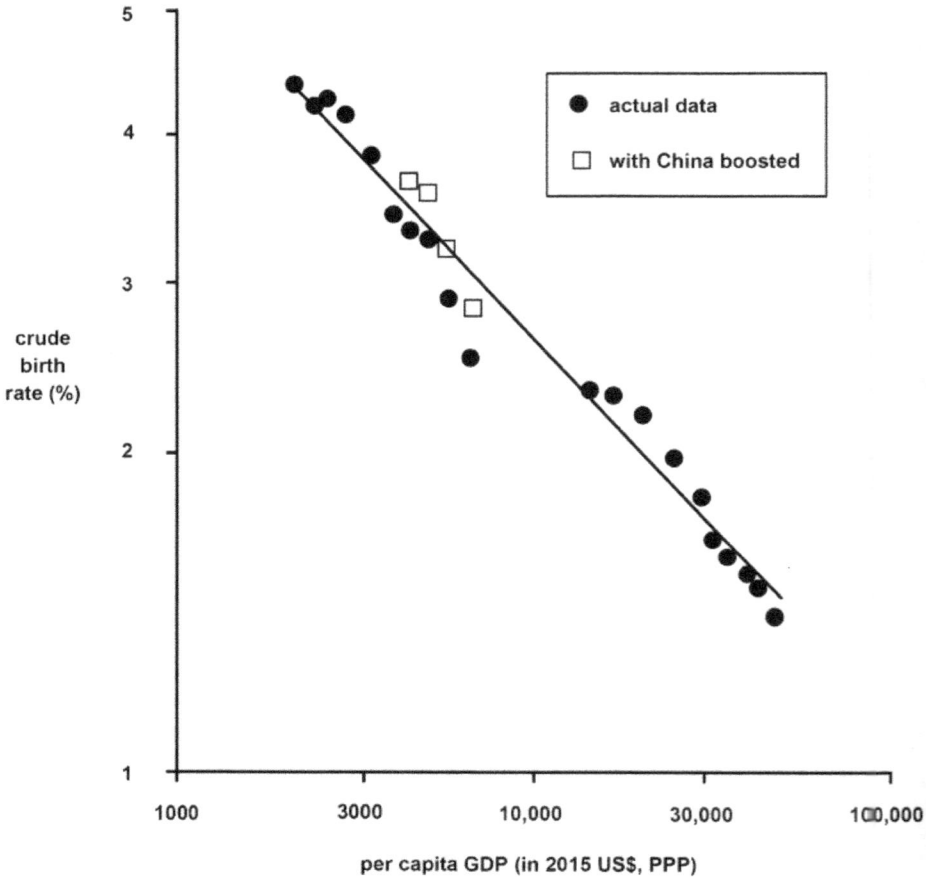

Figure A10. China makes its mark.

China's one-child policy may have prevented the births of 12 million babies per year between 1980 and 2000. This creates only a slight shift in the relationship between per capita GDP and fertility, and the two regression lines are so similar that they overlap. Still, China's policy may have caused a substantial deviation of the Low Income block of nations, below their expected birth rates. The samples and sources are those described in Appendix One. Note the log scales.

Topic H: Residual effects of time on fertility, when prosperity is removed as a factor.

An inspection of Figure A2 in Appendix One shows some evidence of patterns in the data that are not accounted for by the power function regression between per capita GDP and fertility. Figure A11 shows the residual scatter of the data points when the regression equation has been subtracted from their values. The subtracted value at each point is the formula given in Appendix One:

$$\text{fertility} = 291 \ (\text{per capita GDP})^{-0.381}$$

The residual data have been re-plotted against time, and the High Income and Low Income nations are shown separately. Residuals are presented as a percentage of the observed data value.

Although per capita GDP explains 95% of the distribution of fertility rates among the sampling units, the residual scatter that is *not* explained by prosperity can account for up to 27% of the observed data value in specific cases. The residuals seem to be strongly correlated with time, though the real mathematical relationship is not obvious. Still, a glance at Figure A11 shows us that, in both the High Income and Low Income nations of the world, there was a tendency for women to exceed the fertility expectations given by their society's prosperity during the 1950s and 1960s. Since then, this residual fertility factor has been falling sharply.

It's certainly possible that these digressions might be meaningless fluctuations, or part of a long-term cycle of some sort, or otherwise irrelevant to the considerations at hand. But because the trend from 1965 to 2000 is fairly dramatic and is seen worldwide, we must also consider the following possibility. Chapter Nine pointed out that the reason authoritative, mandated solutions are not going to be able to stop the progressive decreasing of fertility is because the internal, neurological drive to carry out such corrective actions is decreasing, worldwide. That process may already be occurring to a degree that is measurable, even in a global dataset like this one. Certainly some examples exist. The examples of the

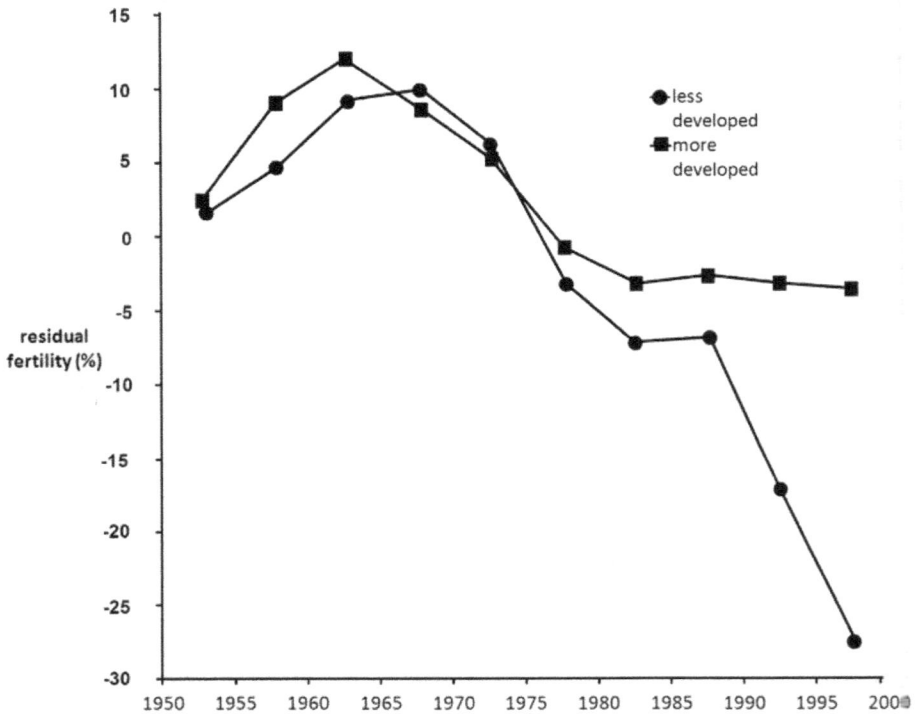

Figure A11. More bad news?

These data show the residual scatter of the data points from
Appendix One, Figure A2, when the regression equation (a
power function) has been subtracted from their values. The
residual data have been re-plotted against time. There seems to
be a progressive, unexplained trend toward falling fertility that
goes beyond even satisfaction sterility. The samples and sources
are those described in Appendix One.

dramatic disobedience of Poland and of Italy to the Catholic prohibition of birth control methods since 1968 was discussed in Chapter Eight.

Note that Figure A11 does *not* show a baby boom in the post-World War Two years caused by an increase in prosperity. The information in the residuals is exactly that information which cannot be explained in terms of prosperity. It would be sensible to wonder why our species seems to be losing fertility even faster than the grim trend of satisfaction sterility can explain. But remember: since about 1980, both China and India have been using legal force to reduce their fertility. Making some tentative corrections for China's one-child policy, as described in Topic G (above) weakens the observed trend, though it doesn't remove it. For example, the most extreme residual, after those adjustments, is reduced to about 17% rather than 27%.

NOTES

Page 1, "It was Thomas Malthus..." Malthus made this observation in his 1798 *Essay on the Principle of Population,* still in print with Oxford World's Classics. A couple of examples of Malthusian revivals can be seen in: John M. Keynes, 1914, *Keynes's manuscript on population;* Pp. 44-74 in: J. Toye, ed. 2000, *Keynes on Population,* Oxford University Press, Oxford UK; and: Paul R. Ehrlich, 1968, *The Population Bomb,* Buccaneer Books, Cutchogue NY. These two Malthus revivals are almost opposite in their concerns. Keynes, who was British, was worried about immigration to the UK from less developed nations, and said: "Almost any measures seem to me to be justified in order to protect our standard of life from injury at the hands of more prolific races. Some definite parceling out of the world may well become necessary." Ehrlich, by contrast, was mainly worried about mass famine creating a humanitarian catastrophe in the Third World.

Page 2, "Nonetheless, incredibly, the vast famines..." An assessment of global food resources in recent generations is found in the UN Food and

Agriculture Organization's 2014 report, *The State of Food Insecurity in the World.*

Page 4, "So I'll be using generations as units of time..." A summary and assessment of the treatment of human generation length is found in: J. N. Fenner, 2005, *Cross-cultural estimation of the human generation interval for use in genetics-based population divergence studies,* American Journal of Physical Anthropology, 128: 415–423.

Page 5, "The second category of supernatural explanation..." Plato discussed his "Theory of Forms" in many of his works, often attributing the theory to Socrates. The theory claims that ideal forms are primary, and that perceived (physical) things are secondary, implying that the former somehow drive or cause the latter. This causative property of ideals becomes especially clear in the Allegory of the Cave, found in Book VII of *The Republic.*

Plato's view was critiqued strongly by Aristotle, who introduced an alternative view, known today as hylomorphism, which suggests that form cannot exist as some sort of pure ideal, separate from matter. I don't take a stance one way or another on hylomorphism, but as far as this book is concerned, we will hold firmly to the following assertion: Even if (in some sense) form *can* exist separate from matter – e.g., in some world of dreams or imagination – it is the result of material processes, such as the workings of the human brain.

Page 6, "Similarly, we'll ignore the idea of Fate..." Although I think Nietzsche meant something supernatural when he used the phrase "Will to Power," I also think he himself may not have actually believed in what he was proposing. He hinted that he was outright lying about it, and merely thought that it would be a healthy thing to believe in, if anyone could bring themselves to do so. He often scoffed at the value of truth, for example when he wrote, "The falseness of a judgement is to us not necessarily an objection to a judgement," on page 35 of: Friedrich Nietzsche, 1886, *Beyond Good and Evil*, Penguin Classics, London. He even proposed that people should (in the interests of good mental health) struggle to delude themselves into believing that their current state was entirely the result of their own previous willful actions; for example: "...to transform every 'It was' into an 'I wanted it thus!' – that alone do I call redemption!" on page 161 of : Friedrich Nietzsche, 1885, *Thus Spoke Zarathustra*, Penguin Classics, London.

I'm not going to argue the point – maybe various forms of intentional self-delusion really do constitute healthy thinking. But, as in the case of religious views, this book is just going to shrug the issue off and carry on with the simpler assumption that efficient natural explanations will suffice for our immediate purposes.

Page 7, "If, on the other hand, you're a reader who is not sure..." A general introduction to the US eugenics programs, especially those of the early 20th century, can be found in: Edwin Black, 2003, *War Against the Weak: Eugenics and America's Campaign to Create a Master Race*, Dialogue Press, Westport CT. An assertion that Oregon's eugenics program ran until 1981 was published in the Portland Oregonian on 11/15/2002, in an article entitled: *State will admit sterilization past.*

Page 7, "The policy spread from the US to many other nations..." Despite the name, "social Darwinism" was invented by the philosopher Herbert Spencer, two years before Darwin's evolution theory. Spencer's *Progress: Its Law and Cause* was published in 1857; Darwin's *Origin of Species* in 1859. Darwin was eventually swayed by Spencer's arguments, and advocated social Darwinism in his 1882 book, *The Descent of Man*. E.g., in Chapter Five: "...[C]are wrongly directed leads to the degeneration of a domestic [species]; but excepting in the case of man himself, hardly anyone is so ignorant as to allow his worst animals to breed."

Some readers may be a little unclear on why 'social Darwinism' is not in keeping with the principles of biological evolution (even if Darwin himself advocated it). Here's the matter in a nutshell. Darwin described a process called 'natural selection', in which the environment of a population selects which individuals live and reproduce, based on circumstantial (but real) criteria. He proposed this mechanism in direct contrast with what farmers do, which is 'artificial selection' – the act of hand-picking which members of a population of plants or animals get to live and breed. We carry out artificial selection programs to make fatter chickens and taller corn. When we decide that a human population 'ought' to have a higher IQ, or a more muscular physique, or a warlike, dominating disposition, then we are working directly against the evolutionary principle of natural selection in order to indulge in artificial selection. We're trying to cultivate our fellow citizens like farm animals.

Page 12, "While we're in this reflective mood..." I'm taking some dramatic liberties, here, with regard to the date. The date of Edison's discovery of the light bulb is apocryphal. Invention is usually the result of a series of innovations, not just one, as the inventors tinker to get their gizmo to work correctly. Edison and his lab staff "invented" the light bulb over and over, in a long, slow series of improvements during late 1879. Nothing special actually happened on October 21; the date was selected arbitrarily as "Edison Day" by Edison Corporation's public relations department, years after the fact. The real story is given in: Robert Friedel and Paul Israel, 1986, *Edison's Electric Light: Biography of an Invention,* Rutgers University Press, New Brunswick NJ.

Page 12, "Have a look at the book..." Regarding Gutenberg: I'm dramatizing again. Moveable type printing was actually invented by Bi Sheng in 1045. That and other innovations from the amazing Song Dynasty are described in: Joseph Needham, 1994, *The Shorter Science and Civilisation in China,* Cambridge University Press, Cambridge UK.

Page 13, "It's impossible to count..." The granting of US patent #9,000,000 was described in the Official Gazette of the US Patent and Trademark Office, 4/7/2015.

Page 15, "If we journeyed back..." It is a matter of debate whether *Attercopus fimbriunguis,* a fossil arachnid from 386 million years ago, built webs. It had spinnerets, but is otherwise not a likely candidate. There's no evidence of anything earlier that might have been a web-builder.

Page 17, "Turgot's story is teleological..." Turgot's speech is still (barely) in print in English translation: A. R. J. Turgot, 1750, *A Philosophical Review of the Successive Advances of the Human Mind. (Tableau philosophique des progrès succcessifs de l'esprit humain)*; collected in: D. Gordon, ed., 2011, *The Turgot Collection*, private publication. The modern term for the kind of pep talk that Turgot gave at the Sorbonne is a "grand narrative" or a "metanarrative." For a critique of this sort of thinking, see: Jean-François Lyotard, 1979, *The Postmodern Condition: A Report on Knowledge. (La condition postmoderne: rapport sur le savoir)*, University of Minnesota Press. People like me who propose inevitable historical trends should, quite rightly, be suspected of trying to foist off some sort of metanarrative on the reader. However, the fact that my story ends with the extinction of our entire species probably gets me off the hook. What kind of manifest destiny is *that*?

Page 17, "Similar teleological feelings of cultural purpose..." The term "manifest destiny" was first used by John O'Sullivan in 1845 to encourage the US government to invade and occupy the Mexican territory of Texas. O'Sullivan's message was explicitly teleological. Although the war had not yet begun, he wrote, "Texas is now ours." He published these views in: J. O'Sullivan, 1845, *Annexation*, United States Magazine and Democratic Review, 17 (1): 5-10.

Page 17, "True, a few cases of non-genetic, intelligence-based..." Food washing in Japanese macaques is reviewed in: M. Nakamichi *et al.*, 1998, *Carrying and Washing of Grass Roots by Free-Ranging Japanese Macaques at Katsuyama*, Folia Primatologica, 69: 35–40.

Page 23, "The difference between 'skilled'..." To license a surgeon, the American Board of Surgery requires a minimum of 5 years of residency, following the usual 12 years of primary and secondary education, 4 years of college, and 4 years of medical school. The US Bureau of Labor Statistics estimates mean annual income for surgeons at $240,440 in 2015. Minimum wage in the US is $7.25 per hour, or roughly $15,000 per year.

Not all differences in pay scale are due to labor value, in the real world. If we look back at the feudal and aristocratic societies, the biggest indicator of the hourly value of a worker's time wasn't so much the complexity of his or her skill set; the big issue was which social class he or she was born into – a situation that hasn't entirely vanished. Other factors that influence labor value are cronyism and nepotism, as well as special talents. The economist Schumpeter provided a list of the latter, which he called the "natural differences in quality" of labor, and his list was: intelligence, willpower, physical strength and agility (Joseph Schumpeter, 1942, *Capitalism, Socialism and Democracy,* Harper Perennial, NY; footnote to p. 24). I think perhaps he left out some important cases, such as good looks (if you're going to Hollywood), risk tolerance (if you're going to Afghanistan), or ruthlessness (if you're going to Wall Street).

Page 26, "At some point during that progressive accumulation..." The definition of civilization as "a developed or advanced state of human society" is from: *Oxford English Dictionary,* 2nd edition, 1991, Oxford University Press.

As mentioned in the text, a lot of anthropologists will find my definition of civilization unduly restrictive. The problem with agreeing to any definition of the word "civilized" in anthropology comes from the pejorative use of the term "uncivilized" (along with "barbarian," "savage," and "primitive") by anthropologists back in the old days. By the mid 20th century, many of the complex social traits that were once believed to exist only in city-based cultures had also been found in city-less societies, especially large chiefdoms. A strong movement appeared in the field to refer to all cultures that showed any such traits "civilized," and it was clear that a lot of the motive for doing so was to free the discourse of anthropology from the culture-bound smugness of its past. I'm all for that. Still, I think it would be a shame to lose the plain clarity of a definition like: "Any culture that contains cities is called a civilization."

Page 26, "Looking back over time, we see that cities..." Regarding the last sentence of this paragraph: Spain's first cities were built by Roman invaders. If we go back further, we find that the Roman civilization first appeared on the Italian Peninsula, where the first cities had been those of Magna Graecia, consisting of Greek colonies. Those colonies were part of Greece's second civilization (after the failed Mycenaean civilization), which spread west across the Adriatic from the Ionian coast, around 800 BC. That coast had for many centuries been part of the loosely bound Phoenician culture (which gave the Greeks the basis of their alphabet and so on). The Phoenicians were among the many Middle Eastern civilizations that arose in the wake of the world's first empire, the Akkadians of Mesopotamia, after the latter's collapse around 2150 BC. No one can really prove that the idea of building cities was introduced to Phoenicia from Mesopotamia, via the Assyrians, as I recklessly claim. But we can say this: the direct successors to the Akkadian civilization *were* Babylonia and Assyria, and of those two, Assyria was the one located right smack between Mesopotamia and the Levant coast – which was Phoenicia's homeland.

Page 27, Figure 2. Because the absolute origin of agriculture is often hard to specify in a given region, I've drawn the bars in Figure 2 using the earliest known *agricultural village* in each region as the "beginning" of agriculture, even if scattered signs of agriculture are older.

For example, Mesopotamia began practicing agriculture 12,500 years ago (D. Zohary and M. Hopf, 2000, *Domestication of Plants in the Old World*, third edition, Oxford University Press) but the region's first known agricultural villages, such as Jarmo, appeared only 9000 years ago (L. S. Braidwood *et al.*, 1983, *Prehistoric Archeology Along the Zagros Flanks,* Oriental Institute Publications 105). Mesopotamia's oldest cities, such as Uruk, formed 6000 years ago (P. Charvát, *et al.*, 2002, *Mesopotamia Before History,* Routledge Press, London).

The Indus Valley's oldest agricultural villages (e.g. Mehrgarh) appeared 8000 years ago (M. Sharif and B. K. Thapar, 1999, *Food-producing communities in Pakistan and Northern India;* pp. 128–137 in: V. M. Masson, *History of civilizations of Central Asia,* Motilal Banarsidass Publisher). The region's first cities (e.g. Mohenjo-daro) formed 4500 years ago (J. M. Kenoyer, 1993, *Ancient Cities of the Indus Valley Civilization,* American Institute of Pakistan Studies, Islamabad). The question of whether the Indus Valley's Harappa civilization really "vanished" or whether it left behind cities that spanned the full period leading to the Mauryan civilization, over 1500 years later, is explored in: A. Lawler, 2008, *Indus Collapse: The End or the Beginning of an Asian Culture?,* Science, 320 (5881): 1281-1283.

Egypt's first agricultural villages (e.g. Faiyum) appeared 7000 years ago (S. Stanek, 2/12/2008, *Egypt's Earliest Farming Village Found,* National Geographic Magazine), and the region's first known cities (e.g. Memphis) formed 4500 years ago (K. A. Bard, 1999, *Encyclopedia of the Archaeology of Ancient Egypt,* Routledge, London). However, there is plausible evidence of undiscovered cities (e.g. Thinis) a thousand years earlier, as described in: D. C. Patch, 1991, *The origin and early development of urbanism in ancient Egypt: A regional study,* dissertation, University of Pennsylvania.

China developed agriculture 10,500 years ago (Xiaoyan Yang *et al.,* 2012, *Early millet use in northern China,* Proceedings of the National Academy of Science, 109 (10): 3726–3730), but the first known agricultural villages, such as Jiahu, arose just 9000 years ago (Xiaoyan Yang *et al.,* 2005, *TL and IRSL dating of Jiahu relics and sediments: clues of 7th millennium BC civilization in central China,* Journal of Archaeological Science, 32 (7): 1045–1051). China's first known cities (e.g. Erlitou) formed 4000 years ago (J. K. Fairbank, M. Goldman, 2006, *China: A New History,* second edition, Belknap Press, Cambridge, MA).

The oldest agricultural villages in the Andes region, such as Real Alto in Valdivia, formed 5500 years ago (D. M. Pearsall, 1988, *An overview of Formative period subsistence in Ecuador: paleoethnobotanical data and perspectives;* pp. 149-164 in: *Diet and Subsistence: Current Archaeological Perspectives,* B. V. Kennedy and G. M. LeMoine, eds., University of Calgary Press). The first known cities in the region (e.g. Caral) formed 4000 years ago (R. S. Solis *et al.,* 2001, *Dating Caral, a preceramic site in the Supe Valley on the central coast of Peru,* Science, 292 (5517): 723-726).

There is evidence of agriculture in Mesoamerica as far back as 10,000 years ago (B. D. Smith, 2001, *Documenting plant domestication: The consilience of biological and archaeological approaches,* Proceedings of the National Academy of Sciences, 98 (4), 1324–1326), but the first known agricultural villages in the region, such as San Andrès, appeared only 4000 years ago (M. Pohl *et al.,* 2004, *Olmec civilization at San Andrès, Tabasco, Mexico,* Foundation for the Advancement of Mesoamerican Studies). The first known cities in the region, such as San Lorenzo Tenochtitlán, formed 3000 years ago (R. Diehl, 2004, *The Olmecs: America's First Civilization,* Thames & Hudson Ltd, London).

Eastern North America's first agricultural villages (e.g., the Martin Farm village) formed about 1000 years ago (T. R. Pauketat, 2005, *The forgotten history of the Mississippians;* pp. 187-211 in: *North American Archaeology,* T. R. Pauketat and D. D. Loren, eds., Blackwell Publishing). The first cities, such as Cahokia, formed in the region 800 years ago (T. R. Pauketat, 1994, *The Ascent of Chiefs: Cahokia and Mississippian Politics in Native North America,* University of Alabama Press). Within a century or so, this brief civilization collapsed.

I've left the African civilization center of Mapungubwe and Great Zimbabwe off my list of seven original civilizations because of its links with the Arab culture found on the Swahili Coast (Kilwa-Kisiwani and other entrepots). Although Mapungubwe achieved civilization as early as the 11th century, and was hundreds of miles from the Swahili Coast, its economy was driven by the gold and ivory trade with the Arab world, and the trade routes to the coast were evidently very active. The case isn't entirely clear, but I'm making the conservative assumption that this was a secondary civilization, inspired by Arab contact. Otherwise... I guess that makes eight. For more information, see: John Stewart, 1989, *African States and Rulers,* McFarland & Company, Jefferson NC.

Page 28, "Of course, one expeditious thing..." Early 20th century anthropologists were the first to become disillusioned about the categorical validity of direct comparison between existing "primitive" societies and our own ancient ancestors. This disillusionment was essentially a collapse of the teleological view of progress that Turgot had first expounded. From 1750 to about 1910, almost every anthropologist subscribed to the notion that cultures advance up a single ladder, starting as barbarous primitives and (if lucky) advancing to the status of Christian republics endowed with industrial wealth. That bubble was popped by Franz Boas, a German-American anthropologist who introduced cultural relativism to the field in the early decades of the 20th century.

Page 29, "Certainly, one thing that's intriguing..." At a casual glance, it might seem that the large insect societies such as those of ants, bees or termites show strong parallels with modern industrial civilizations. They are populous, well organized, often have distinct military and productive sectors, and build synthetic environments that have a lot of resemblance to miniature cities. But when we try to extend the comparison to include biological and genetic factors, it falls apart. An ant hill, bee hive or termite colony is actually a single, immense family unit... it's really more like an oversized hunter-gatherer band than it is like a republic. The intense social relations within a hive are easily explained by evolutionary science because of the strong kin relations. Big human societies are more challenging to explain through biological reasoning, and have so far resisted all efforts to do so.

Page 29, "A third thing to notice about the band cultures..." Regarding the assertion that mass storage is rare in band societies, note that nonetheless granaries seem to have preceded agriculture in some regions, indicating that grain accumulation was important in some cases even before methods of grain cultivation existed. This discovery was reported in: I. Kuijt and E. Finlayson, 2009, *Evidence for food storage and predomestication granaries 11,000 years ago in the Jordan Valley*, Proceedings of the National Academy of Science, 106 (27): 10966–10970. Though interesting, the discovery isn't entirely surprising. Various non-agricultural societies stored dried fish, smoked meat and other winter supplies, and of course many nonhuman animals also store food. There are even non-human species with social systems that are heavily structured around the requirements of their food storage system (e.g., acorn woodpeckers and honeybees).

Humans have been innovating since before our species existed. Our first known innovations, knapped stone tools, were made by our ancestral species, *Homo erectus*, about 2.6 million years ago, and comprise the Oldowan toolkit from Ethiopia. See: S. Semaw *et al.*, 2003, *2.6-Million-year-old stone tools and associated bones from OGS-6 and OGS-7, Gona, Afar, Ethiopia*, Journal of Human Evolution 45 (2): 169–177. Chimpanzees also manufacture tools, though not out of stone, so perhaps a more definitive landmark in the history of human innovation was the discovery of ways to make fire. Again, this innovation was probably not made by us *Homo sapiens* but by an ancestral species. The date of the oldest campfire is contested, with estimates ranging from 200,000 to 1.8 million years ago. See: S. R. James, 1989, *Hominid use of fire in the Lower and Middle Pleistocene: a review of the evidence*, Current Anthropology 30 (1): 1–26. This reference is not up to date, but the discoveries that have turned up since that time (though interesting) haven't settled the matter.

Page 30, "At each of the seven origins of civilization..." Anyone who wants to delve into the vast and contentious literature on the origins of agriculture might start with: C. Wesley Cowan and Patty Jo Watson, 1992, *The Origins of Agriculture: An International Perspective*. Smithsonian Institution Press, Washington DC.

Page 32, "I leave it to the reader..." Here is a quick tour of some of the leading lights in the 150-year squabble over the nature of cultural progress, to help the interested reader get started.

During the generation of Nietzsche, more than a century after Turgot's speech, two key developments happened in the formation of the new science of sociology. The first was the 1873 publication of Herbert Spencer's book, *The Study of Sociology*. Spencer noted that societies, like organisms, must struggle against each other in order to survive. He felt that they advance to more complex states due to an "evolutionary force" that is directly analogous to biological evolution. Although modern sociologists and modern evolutionary biologists have many disagreements, most of us are united in seeing Spencer's views as woefully wrong-headed. Still, they were very influential at the time.

Working around the same time as Spencer, Karl Marx felt that the driving force leading to large, complex states was the conquest and exploitation of laborers by militarily empowered bullies. As the process of exploitation becomes more refined, these bullies evolve from slavers into aristocrats and then to capitalists. Marx developed his views on the impositional nature of the state throughout his career, but his most trenchant observations are found in the *Grundisse*, which was first fully translated into English in 1973, ninety years after his death. See: Karl Marx, 1973, *Grundisse (Outlines of the Critique of Political Economy)*, Penguin Classics, London.

Two generations after Spencer and Marx, Max Weber attempted to expand the understanding of societies and their developmental processes by considering the intents of the people involved, rather than simply observing the resulting structures – an "antipositivist" approach. This put a big damper on the naive, 19th century efforts to apply Darwinian principles to human cultures. Weber ended up agreeing with Marx about the basic relationship of government and governed, saying: "the state is that entity which claims the monopoly of the legitimate use of force in a given region" (Max Weber, 1919, *Politics as a vocation.* In: *The Vocation Lectures.* Hackett Classics, Indianapolis IN).

After another generation passed, V. Gordon Childe was the first to really convince a lot of anthropologists, historians and sociologists that the innovations of agriculture and of cities were key points in the development of complex human societies. In his 1942 book, *What Happened in History*, he referred to the first of those two events as the "Neolithic Revolution," and the second as the "Urban Revolution," and those terms are still in common use. (Incidentally, given his essentially "evolutionary" views, it is perhaps ironic that Childe was a vehement Marxist.)

Around the same time as Childe, Julian Steward was a major leader of the Neoevolutionary movement in anthropology. The movement began in the 1930s as an effort to get some use out of Darwinian principles despite the antipositivist critiques of Weber and others. The movement emphasized the idea that societies, like evolving lineages of organisms, adapt to their environments. These views are perhaps best expressed in:

Julian Steward, 1955, *Theory of Culture Change: The Methodology of Multilinear Evolution*, University of Illinois Press. I have to add here that any evolutionary biologist would wince at such comparisons, which seem hopelessly naive.

In the next generation, one of Steward's ex-students, Elman Service, rebelled strongly against the notion of evolutionary causality in the development of complex societies. Unlike Spencer, Marx and Steward, Service made a real effort *not* to claim that he could see some underlying drive that caused the formation of civilizations; rather, he just elucidated the evident patterns. His role in quashing Neoevolutionism may have been similar to Weber's role in quashing the 19th century social evolution movement. See: Elman Service, 1975, *Origins of the State and Civilization: the Process of Cultural Evolution,* W.W. Norton, NY.

A third effort was made to muster interest in the possibility of biological evolutionary processes as the basis of cultural complexity, starting in the mid-1970s. This movement, called "sociobiology," was spearheaded by evolutionary biologist Edward Wilson, who proposed that cultural development is based on innate tendencies that have been genetically programmed into the human brain by natural selection (especially kin selection). He expressed these views in: Edward O. Wilson, 1975, *Sociobiology: The New Synthesis*, Harvard University Press. See especially pp. 573-574 of that book, where Wilson's argument becomes painfully close to that of Spencer. Though Wilson's views were (and are) more biologically sophisticated than those of Spencer or Steward, they proved nearly useless in explaining anything interesting about human anthropology or sociology. By the mid-1980s, interest in the 'sociobiology' approach dwindled away to nearly nothing.

Page 32, "Next question: What do we mean by 'city'?" Regarding Carthage, for the reader who doesn't have instant access to those hazy memories from high school lectures: the Romans destroyed Carthage in 146 BC, sold all 50,000 inhabitants into slavery, and disassembled every building stone by stone. When nothing was left but empty fields, they reportedly sowed the fields with salt.

Page 33, "There's a good reason that..." Brendan O'Flaherty reinforces the concept that wall fortifications were the essence of the first cities, and attempts to give a mathematical analysis of the economics that led to the conditions in which walls could first be afforded by a proto-city. See: Brendan O'Flaherty, 2005, *City Economics*, Harvard University Press.

Note, though, that V. Gordon Childe left us a famous list of ten essential qualities that define a city, and this list is still considered one of the best descriptive tools for figuring out what is and what is not a city. The list first appeared in (of all places): V. G. Childe, 1950, *The Urban Revolution*, Town Planning Review, 21 (1): 3–17. The ten factors include such things as high population density, tax, specialized professions, and monumental architecture, but *not* protective walls. Rightly so: fortifications are not seen surrounding most modern cities. Fortifications aren't characteristic of all cities, but they do distinguish the first cities from the proto-cities that came before them.

Page 36, "The fact that innovations can pile up..." Schumpeter's feelings about progress were very nuanced. He was a product of his economic times, which were dominated in the US by the 1920s bubble economy, the Depression, and the sluggish recovery during the New Deal years. He is mainly remembered for proposing that capitalism is cyclic, following a pattern of "creative destruction" in which new industries suck up all the investment capital and cause the old industries to wither or crash. Nonetheless, he was consistent in saying that innovation is the drive of economic growth. See, for example, Chapter 10 of: Joseph Schumpeter, 1942, *Capitalism, Socialism and Democracy*, Harper Perennial, NY.

Page 37, "At the other extreme, Karl Marx's views..." Marx's comment about windmills (sometimes translated "hand-mills") and steam-mills is from page 119 of: Karl Marx, 1847, *The Poverty of Philosophy*, Charles H. Kerr & Co., Chicago.

Page 37, "The use of the term 'developed'..." The terms "more developed" and "less developed," commonly used by the UN and other organizations to compare nations, are rapidly losing this carefully crafted non-teleological quality. In 1990, economists Amartya Sen and Mahbub ul Haq introduced the HDI, or Human Development Index, explicitly linking the term "development" to teleological progress toward humanitarian goals. The UN adopted the HDI, and it remains in common use.

Page 38, "For the purposes of this book, per capita GDP..." More rigorously, when I say that per capita GDP is a strong indicator of the popular experience of prosperity, what I'm saying is this: A) To a first approximation, productivity is the result of labor. B) To a first approximation, income is fair compensation for labor. C) To the degree that those two statements are sufficiently close approximations to reality, a nation's per capita GDP is a good indicator of the average person's experience of prosperity or poverty in that nation.

Page 39, "The difference between a nation's GDP..." In 2013, Malaysia's GDP was $313 billion, with per capita PPP value of $23,300, while Singapore's GDP was $298 billion, with a per capita PPP value of $78,800. Data from: *World Development Indicators*, World Bank, 2015.

Page 40, "At the end of each week..." In the US in 2013, residential building construction workers made an average of $32,850, according to the US Bureau of Labor Statistics. The salary of an Indonesian construction worker is about 2.5 million rupiah per month, which is $2280 per year by the official exchange rate. The money goes quite a bit further than that on the local economy (adjusted to purchasing power parity), so the Indonesian worker's real level of prosperity is that of a person living in the US on an annual income of $3950. (PPP exchange value from: *World Development Indicators*, World Bank, 2015.)

Page 43, "As prosperity begins to rise in a less developed..." Regarding the last sentence of this paragraph: In 1841, Ireland's population was 8.2 million, and it fell to 6.6 million by 1851, according to census data reported in: Cecil Woodham-Smith, 1962, *The Great Hunger: Ireland 1845–1849*, Penguin, London. The proximate causes of Ireland's famine were complicated (one of them being the potato blight fungus, and another being some conspicuous social injustices), but it's clear that the island was populated beyond the level of sustainability. The surest indicator of this is the fact that Ireland's population has still never returned to pre-famine levels. According to the World Bank, the island as a whole had 6.4 million population in 2013 – the highest on record in over 150 years.

Page 49, "So all motivational drives..." Freud described the pleasure principle as the sole guiding drive behind the id, while the mature ego is guided by the reality principle. See: Sigmund Freud, 1911, *Formulations regarding the two principles of mental functioning;* pp 13-21 in: 1950, *Collected Papers of Sigmund Freud*, volume 4, Hogarth Press and The Institute of Psychoanalysis, London.

Page 51, "Now we should dig a little..." The physiologically savvy reader will notice that throughout the following exegesis of basic neurophysiology, I've struggled to avoid the dense jargon of the field. I hope it's clear from context that by "brain cells" I mean neurons rather than astrocytes, microglia and so on; by "extensions" I mean axons and telodendria rather than dendrites; by "signal molecules" I mean neurotransmitters rather than hormones or cytokines; and by "vesicles" I mean specifically the synaptic vesicles of the axonal terminal. For readers who don't know what all that stuff means, don't worry... I think the gist will come through clearly enough.

Page 52, "Only the simplest examples of these neurologically based..." Regarding the last sentence of this paragraph: The muscular reflexes of the limbs are mediated through the spinal cord, heart rhythm is myogenic, and some digestive reflexes are mediated through the plexus of enteric neurons. All the other muscular reflexes are integrated in the brain, either through the autonomic nervous system (e.g. vasoconstriction in a cold limb) or various gray matter nuclei (e.g. the startle reaction to a sudden noise).

Page 52, "Hunger works like this." I've left most of the hormonal interactions out of this story for brevity. Here are a few of the more interesting ones. In addition to parasympathetic vagus stimulation of the brain by the stomach, prolonged absence of food in the gastrointestinal tract causes the pancreas to dump glucagon and epinephrine into the blood. Both of these hormones circulate to the brain and stimulate hunger. The stomach itself produces the hormone ghrelin when the brain hints that there's a meal coming up, and the ghrelin in turn stimulates the hunger center. Meanwhile, the stomach self-stimulates with gastrin during this appetitive excitation phase. There are also some remarkable inhibitory feedbacks. For example, the vagus nerve carries inhibitory signals to the brain that not only report that the stomach is full but even specify the categories of nutrient that it contains. Presumably, that's a major basis of our feelings of being hungry for one type of food rather than another. Another remarkable appetite inhibition system is leptin, a hormone released by adipose (fat storage) cells all over the body in order to decrease our appetites if we're getting overweight. Low leptin production is a genetic variant that predisposes some people to morbid obesity.

The key roles of orexin and ghrelin in modulating hunger in the lateral hypothalamus are reviewed in: Jon F. Davis *et al.*, 2011, *Orexigenic hypothalamic peptides, behavior and feeding*, pp. 355-370 in Victor R. Preedy *et al.* *Handbook of Behavior, Food and Nutrition*, Springer NY.

Page 54, "Separate from the pleasure..." The role of the peptide neurotransmitters substance P and neurokinin A in stimulating pain circuits in the brain was first described in: Y. Q. Cao *et al*, 1998, *Primary afferent tachykinins are required to experience moderate to intense pain*, Nature, 392: 390-394. The dorsal posterior insula was identified as the (or at least a) key pain center in: A. R. Segerdahl *et al.*, 2015, *The dorsal posterior insula subserves a fundamental role in human pain*, Nature Neuroscience, 18: 499–500.

Page 57, "The strongest voice..." Nietzsche introduced the phrase 'will to power' in *Thus Spoke Zarathustra* (1883), immediately contrasting it against the pleasure principle: "What it accounts hard, it calls praiseworthy," (p 34 in the Penguin Classics edition). He went on to assert that there is no such thing as a will to simply exist (p. 138), and to claim that living things are motivated by the more dynamic will to power. It is worth noticing, however, that he later hints that his reason for expressing these views may not necessarily be that he feels them to be true, but rather that he feels them to be a healthy, robust set of beliefs (pp. 162-163).

There are other views on the matter. Max Weber, in his 1905 book, *The Protestant Ethic and the Spirit of Capitalism* (Penguin Twentieth-Century Classics, London), proposed that an attitude like that of the Third Brewer is at the heart of the capitalist ethos, and can be traced back directly to the self-abnegating, anti-worldly attitudes of Martin Luther, the Puritans, and other ascetic Protestants. Obviously, examples of the syndrome can also be found in pre-Reformation history, but Weber nonetheless had a point.

Page 58, " Incidentally, Nietzsche also wanted..." See: Dirk R. Johnson, 2010, *Nietzsche's Anti-Darwinism,* Cambridge University Press.

Page 66, "That's evolution." When I single out natural selection as more interesting than other modalities of evolution, I'm including sexual selection (not described in this book) as a form of natural selection. That wasn't Darwin's original system of terminology, but it's increasingly common practice. The other modalities of evolution are neutral selection (including the special case of genetic drift), population admixture, and mutation pressure (including the special case of speciation events by polyploidy and other radical mutations).

Page 68, "The driving 'force' behind..." More precisely, fitness is the capacity of an individual to pass on its DNA code to future generations, under a given set of circumstances. That's subtly different from Darwin's criterion of survival and reproduction, as we'll see. Keep in mind that even the most basic facts of genetics were unknown in Darwin's time.

Page 68, "Although the word 'fitness'..." In all fairness, when Spencer introduced the phrase "survival of the fittest" on page 444 of his *Principles of Biology*, he pointed out in the very same sentence that what he really meant was "multiplication of the fittest." This latter phrase, which at least gets somewhat closer to real Darwinian thinking, didn't catch on (predictably enough). Either way, though, I think it's possible that when Spencer used the word 'fittest', he was purposely choosing to express matters in terms that would confuse people for the next 150 years by encouraging them to think that Darwin's views justified efforts at aggressive domination. It was Spencer who invented "social Darwinism" (though, as mentioned earlier, it was Julian Huxley who whipped the idea up to a genocidal level of enthusiasm). Spencer was thus the grandfather of 20th century eugenics, forced sterilization programs, etc. If he understood Darwin's original work at all, he was engaged in an effort to derail Darwin's ideas in the hope of justifying England's ongoing imperial efforts around the world, and bolstering the class system at home. It might strike the reader that this fascination with dominance, conquest and oppression sounds a lot like Nietzsche, but actually Nietzsche despised Spencer, regarding him as an insipid ivory-tower intellectual. Go figure.

Page 68, "The problem with the word 'fitness' is..." The phrase "Nature, red in tooth and claw" is from the late Romantic poet Alfred Tennyson's 1849 poem, *In memoriam A.H.H.* Darwin's book, *The Origin of Species,* came ten years later.

Page 69, " The descriptive term 'fitness'..." In population genetics, absolute fitness (w_{abs}) of a genotype is defined as the number of individuals with that genotype in one generation divided by the number that had it in the previous generation.

Page 69, "Let's go back to our mouse anecdote for a moment." We use the term 'Lamarckism' to describe the idea that innovative traits might be acquired by individual organisms – whether by luck or an act of will – and then passed on genetically to their offspring. This term gives a nod to Jean-Baptiste Lamarck, the French scientist who popularized this false view of evolution, two generations before Darwin. Incidentally, Charles Darwin's grandfather, Erasmus Darwin, was one of Lamarck's earliest and most outspoken fans. Although he died within months of Lamarck's first publication of the theory (1802; *Researches on the Organization of Living Bodies*), Erasmus managed before dying to finish a long poem about it, *The Temple of Nature*, which was published posthumously.

Page 71, "This fact is the basis of..." The phenomenon of nest helping was first reported by Alexander Skutch, without reference to kin selection, in: A. F. Skutch, 1935, *Helpers at the nest*, Auk, 52 (3): 257–273. Since then, the original term 'helper at the nest' has been replaced by the terms 'nest helper' and 'den helper', at least among American ecologists. To read a good review of nest and den helpers, and the application of kin selection to these cases, see: T. Clutton-Brock, 2002, *Breeding together: kin selection and mutualism in cooperative vertebrates*, Science, 296 (5565): 69-72.

Page 72, "Because of this odd situation..." Inclusive fitness (the basis of kin selection) was first introduced by William Hamilton in: W. D. Hamilton, 1964, *The genetical evolution of social behaviour*, Journal of Theoretical Biology, 7 (1): 1–52. Hamilton was especially interested in the application of this view to the hymenopteran insects (ants, bees, wasps), which have an unusual genetic system called haplodiploidy that causes each female to share *more* DNA code with her sisters than with her own daughters. As a result, females actually prefer to stay home and help their mother (the queen) to raise her eggs, rather than leave and have young of their own. This leads to the immense social groups seen in ant and bee colonies.

Page 73, "The motivational circuitry..." Obviously, we can't really be certain how the neuronal circuitry of the brain was rigged up back in 10,000 BC, because neurons don't leave fossils behind. What we *can* say (and all I'm really asserting here) is that: A) Cranial capacity in our direct lineage has tripled in the past 4.5 million years (from 450 ml to 1350 ml), but has not changed appreciably in the past 100,000 years. B) A lot of essential neurological capabilities such as 'intelligence' are well correlated with characteristics of the bony cranial cavity across a wide spectrum of extant vertebrates, and these things are *especially* well correlated within narrow phylogenetic groups such as our family, the Hominidae. C) Given A and B, the proposition that our immense behavioral changes since 10,000 BC are the result of a physical re-wiring of our neuronal circuitry is so far-fetched that I don't think any serious scientist has ever even proposed it.

Page 76, "Putting these observations together, we can make..." Regarding the last sentence of this paragraph: Our species, *Homo sapiens,* was preceded (in reverse chronological order) by *Homo heidelbergensis, Homo erectus, Homo ergaster, Australopithecus africanus, Australopithecus afarensis, Ardepithecus ramidus,* and *Sahelanthropus tcahdensis.* That gets you back to seven million years ago, when our ancestors branched away from the chimpanzees and bonobos. But reader, beware: almost every word in the previous two sentences is subject to rabid debate among physical anthropologists, and they (not I) are the experts. The theory-to-data ratio is more top-heavy in their field than in any other region of academia, even archaeology. But, for our purposes, the important thing is just this: Every species I've just listed, including the chimpanzees and bonobos (and us, until 600 generations ago), lives or is believed to have lived in clan groups of at most a few dozen individuals.

Page 78, "Humans, like other mammals..." To be technically concise, we're not really talking about eggs, here. Human reproduction is characterized by fertilization of a pre-egg cell called an öocyte, at a point when female meiosis is still incomplete. There's no such thing as an unfertilized egg in humans, so no human sperm has ever literally "fertilized an egg."

Page 79, "In terms of brain circuitry..." The literature on the neurology of libido is reviewed in: J. G. Pfaus, 2009, *Pathways of sexual desire,* Journal of Sexual Medicine, 6 (6): 1506-1533. Pfaus notes that in addition to the dopaminergic axonal links between the hypothalamus and other limbic system centers, there are also important roles played by melanocortin, oxytocin and norepinephrine.

Page 79, "The first of these is..." The main papers from the first fifteen years of research on pregnancy endorphins are cited in: A. N. Margioris, 1993, *Corticotropin-releasing hormone and pro-opiomelanocortin in placenta and fetal membranes,* pp. 277-289 of G. E. Rice & S. P. Brennecke, eds., *Molecular Aspects of Placental and Fetal Membrane Autacoids,* CRC Press, Florida. Interestingly, although endorphins are pleasure-inducing neurotransmitters, they are libidinal suppressors, or at least suppressors of orgasm.

Page 80, "The most important set..." The discovery that pictures of cute babies causes activity in the nucleus accumbens was reported in: M. L. Glocker *et al.,* 2009, *Baby schema modulates the brain reward system in nulliparous women,* Proceedings of the National Academy of Sciences USA, 106 (22): 9115–9119.

Page 86, "My story about the three..." Specialists in evolutionary ecology will recognize that my term 'post-Darwinian threshold' is actually a form of the well known process called ecological release. In ecological release, a population is relieved of its pre-existing selective pressures, for example during a colonization event or by surviving an ecological shift that kills off the population's predators and/or competing species. The term 'ecological release' was coined by Alan Kohn in: A. J. Kohn, 1972, Conus miliaris *at Easter Island – ecological release of diet and habitat in an isolated population,* American Zoologist, 12: 712.

The phrase "beyond the post-Darwinian threshold" describes any case of ecological release in which selective pressure becomes temporarily negligible due to resource enhancement. Many invasive species pass through periods of this sort, before crowding occurs. Unfortunately, such events are ephemeral, and even the best studied cases don't provide us with a systematic understanding that we could use to gaze upon ourselves. The interested reader might nonetheless have a look at: Daniel Simberloff and Marcel Rejmánek, eds., 2011, *Encyclopedia of Biological Invasions,* University of California Press, Berkeley CA. My assessment of the case of the small Indian mongoose is included in that book, on p. 631. Odd things do happen to animal behavior under such conditions, but we humans may be the first recorded species to *lose* reproductive drive as a result of resource enhancement.

Page 87, "A long time ago..." The estimate of per capita GDP in Holland in 1700, and comparisons with the level of prosperity in other nations in the past and present, as well as the estimate of global per capita GDP in 1945, are from: Angus Maddison, 2003, *The World Economy: Historical Statistics,* OECD.

While some readers may be amazed to realize that we modern humans are outside the reach of natural selection, others may be equally amazed at the notion that most of us were *not* beyond its grasp until 1945. After all, by that time, civilization had been around for thousands of years. But the appearance of civilization didn't immediately pull humanity from the merciless jaws of natural selection. During most of civilization's history, just as in pre-civilized times, the average person in every society lived at a level of prosperity below the post-Darwinian threshold, experiencing natural selection as an immediate, palpable stress. Here's a rule of thumb: If natural selection is part of your life at all, then it consumes almost every scrap of your time and energy. A few examples of the ways in which natural selection presents itself to people, day to day, are the threat of imminent starvation, outbreaks of communicable diseases that threaten to kill your children, the possibility of freezing to death before dawn, and the lurking presence of some man-eating predator. If things like that aren't part of your daily routine, then you probably live above the post-Darwinian threshold.

Page 88, "Since this book is really..." To be more precise, fitness among humans in wealthy nations consists of the tendency of the individual *and close relatives* to have children. But there is no substantial 'nest helper' effect in that extended fitness. If I were a miserly millionaire who found himself presented with five orphaned nephews, and refused to give them a penny or even food and shelter, then my nastiness would have almost no negative effect upon my fitness. The kids would go to a state facility or foster home, survive their miserable childhood, and grow up to be just as reproductively capable as if I had treated them decently. So the difference between inclusive fitness and individual fitness is little more than a technicality among the citizens of developed nations.

Page 90, "Indeed it is." It was the neurophysiologist James Olds who discovered the euphoric and addictive effect of electrical self-stimulation of the nucleus accumbens. See: J. Olds, 1956, *Pleasure center in the brain,* Scientific American, 195 (10): 105-16.

Page 91, "A more familiar example..." Cocaine is a dopamine re-uptake inhibitor. That means that when dopamine is released from an axonal terminal, there's a delay before the terminal sucks the dopamine back up into its vesicles for re-use. That in turn means that the dopamine molecules remain exposed to the target cell for a longer period, creating more pleasure than the synapse was evolutionarily designed to give.

Regarding the fact that it's easy to addict lab rats to cocaine, please note, however, that there have also been many demonstrations that dopaminergic addiction is primed by low-quality environments. For an example of a paper showing that low-quality environments contribute to cocaine addiction in rats, see: E. Bezard *et al.,* 2003, *Enriched Environment Confers Resistance to 1-Methyl-4-Phenyl-1,2,3,6-Tetrahydropyridine and Cocaine: Involvement of Dopamine Transporter and Trophic Factors,* Journal of Neuroscience, 23 (35): 10999-11007.

Page 94, "Artificial, innovative methods of decreasing our dissatisfaction.." The ancestor of domestic cattle was the auroch, *Bos primigenius.* These large wild oxen had a broad range in Eurasia and North Africa, and are widely depicted in cave paintings. The last one is said to have died in Poland in 1627. See: Cis van Vuure, 2005, *Retracing The Aurochs,* Pensoft Publishers, Sofia, Bulgaria.

Page 100, "In addition to contraception..." President Clinton's denial of having "sexual relations" with Monica Lewinsky occurred during a deposition at an unrelated case, dealing with his alleged sexual harassment of Paula Jones. That case was settled out of court, but when forensic evidence of Lewinski's fellatio of the president was brought to Congress, the US House of Representatives impeached him in December 1998, for perjury in the Jones trial.

Page 101, "Biologically, of course, Clinton was..." Sex *is* a biological phenomenon, from one point of view. Biologists think of sexual life as a six-part cycle consisting of meiosis, gametogenesis, intromission, fertilization, syngamy, and multicellular development. Of these six events, only one is also found on the long list of acts that are acknowledged as sex by most modern people. That one is intromission: the transfer of viable sperm cells to a position near the cervix, within swimming distance of a viable oöcyte. Fellatio is certainly not an example of intromission, and is thus not biological sex – but it is nonetheless sex from the viewpoint of anyone but a biologist.

Page 101, "Sigmund Freud was way ahead..." There has been a lot of misunderstanding about the term 'polymorphous perversity'. Freud defined this term to mean a potential consequence of the sexual molestation of children, not as a typical developmental stage or outcome. He introduced the phrase in the following way: "Under the influence of seduction children can become polymorphously perverse, and can be led into all possible kinds of sexual irregularities." See page 57 of: Sigmund Freud, 1905, *Three Essays on the Theory of Sexuality*, Martino Fine Books, Eastford CT.

Page 102, "There are several ways of..." The term 'adoption' is used throughout this discussion to mean the adoption of non-relatives. Adopting a nephew or niece (for example) could be entirely explained in terms of inclusive fitness and kin selection. But those arguments don't apply if the adopted child is completely unrelated to both parents.

Page 103, "The urge to adopt is..." The statistic of 3% in this paragraph is deduced from two sources of raw data. There were 136,000 adoptions in the US in 2008, according to the Child Welfare Information Gateway of the US Department of Health and Human Services, as described in their 2011 publication: *How many children were adopted in 2007 and 2008?* There were 4,250,000 births in the US in 2008, according to: J. Martin *et al*, 2008, *Births: Final Data for 2008,* National Vital Statistics Reports, 59 (1): 1-72. Estimates of the price of US adoption in 2011 can be found in: *Factsheet for families,* 2011, Child Welfare Information Gateway (US Dept. of Health and Human Services).

Page 103, "Since the adoption of unrelated babies..." The data on adoption in penguins is from: P. Jouventin *et al.* 1995. *Adoption in the emperor penguin,* Aptenodytes forsteri. Animal Behaviour, 50(4): 1023-1029. Over two thousand chicks were documented in this study.

Page 104, "Not all cross-species adoption..." The case of Koko and her pet cats was covered widely by the media, over the course of many years. See, for example: C. McGraw, 1/10/1985, *Gorilla's Pets: Koko Mourns Kitten's Death*, Los Angeles Times.

Page 105, " In the wild, on the other hand,..." This case of wild, cross-species adoption was reported, for example, in: BBC News Agency, 1/7/2002, *The lioness and the oryx.*

Page 105, " The comparison between loving..." The observation that we have the same sorts of reactions to human babies and cute pets was formally established in a quantitative behavior study by: J. Golle *et al.*, 2013, *Sweet puppies and cute babies: perceptual adaptation to babyfacedness transfers across species,* PLoS ONE [a peer-reviewed online journal] 3/13/2013.

Page 106, "It may be interesting, in passing..." The word 'cute' appeared in the 1830s as a slang word, and its development is traced in the *Oxford English Dictionary,* 2nd edition, 1991, Oxford University Press. The development was: "He/she looks sharp,"... "He/she looks acute,".. "He/she looks 'cute." As for the later, non-sexual meaning of the word, it's notable that most languages still lack a direct translation of the English word 'cute' in its non-sexual connotation – though Japanese has the word *kawaii*.

Page 106, "Another indication that this might..." The story of the English parliament bursting into laughter is from page 41 of: S. Brooman and D. Legge 1997, *Law Relating to Animals*, Routledge-Cavendish, UK. The California man doing 25-years-to-life is James Abernathy, convicted of felony animal cruelty in 2004 and sentenced under a 'third strike' law, as reported in: C. Luna, 10/9/2004, *Dog beheading case illustrates Prop. 66 battle*, Los Angeles Times.

Page 107, "Statistics show that that's exactly..." The statistic of 3.1% in this paragraph is deduced from the following numbers. Americans spent $17 billion on their pets in 1994, and $58 billion in 2014, according to the US Bureau of Labor Statistics' 2015 publication, *Consumer Expenditure Diary and Interview Surveys*. US population grew 22% in that period according to the US Census Bureau, and the dollar's value fell by 37% as measured by the US Department of Labor's Consumer Price Index. Combining those four raw numbers indicates a 3.1% annual rate of real per capita increase, and a doubling time of 23 years. Statistics on pet expenditures prior to 1994 are less reliable, but suggest that this trend goes back a century or more.

Page 110, "I suppose I should admit that..." The link between high per capita GDP and low fertility is very well established, but some confusion may exist among economists and demographers. This can largely be traced to a paper by statistician Mikko Myrskylä and his colleagues in 2009, entitled *Advances in development reverse fertility declines* (Nature, 460: 741-743). The graphs in that paper could easily be misconstrued by casual readers as showing a reversal in the relationship between economic development and fertility when per capita GDP exceeds some particular level. The unfortunate title of the paper is especially misleading in that regard. In fact, however, the authors never claimed to have spotted any evidence of a weakened or reversed relationship between increasing prosperity and decreasing fertility – not at any level of per capita GDP. They only claimed that the HDI (Human Development Index), a basket of sociological factors that includes education and so forth, is somehow correlated with an increase in fertility *despite* the fact that one of the independent variables in the HDI basket is per capita GDP. The truth is that the high per capita GDP of the high-HDI countries is pulling one way (decreasing fertility), while something else in the basket – some cluster of sociological factors – is creating a pull in the other direction. That's hardly surprising. Chapter Eight of this book is devoted mainly to considering sociological factors that can have that sort of effect, at least for a generation or two at a time.

Page 112, "Here are the numbers..." The statistics in this paragraph are from the UN Population Division's *World Population Prospects: The 2015 Revision*, and the UN's Population Database.

Page 112, "The days are past..." The statistics in this paragraph are from the UN Population Division's 2013 publication, *World Population Ageing*.

Page 117, "Here's what we really know..." The 3-7% involuntary infertility statistic is from: W. Himmel *et. al.*, 1997, *Voluntary childlessness and being childfree*, British Journal of General Practice, 47 (415): 111–118. The scientific review of the many studies showing (or failing to show) declining sperm counts is: H. Fisch, 2008, *Declining worldwide sperm counts: disproving a myth*, Urologic Clinics of North America, 35: 137–146.

Page 119, "Survey interviewers in Europe..." The survey in Japan was reported on page 3 of *The Daily Yomiuri* on April 3, 2003.

Page 129, "The second thing we can say..." The only large, UN-defined developed nations that are growing at or above the world average rate are Norway and Australia, as of 2012... and in both cases, the growth is due to immigration. In fact, 72% of Norway's new citizens in 2012 were due to immigration, not births. Data for the growth rates of nations are from the 2013 *CIA World Factbook*. Data on immigration and birth in Norway are from Statistics Norway.

Page 130, "Unfortunately, biology doesn't back us up on this one." The set of needs that ecologically defines a species is known as its niche. The *competitive exclusion principle* says that two phylogenetically defined species can't occupy the same niche.

Page 132, "Even starvation won't stabilize..." If you're not averse to a bit of straightforward math, you'll find the remarkable phenomenon of fluctuating population dynamics clearly introduced in: Alan Hastings, 1997, *Population Biology: Concepts and Models*, Springer, NY.

Page 132, "A classic example of this..." The details of this case come from: V. B. Scheffer, 1951, *The rise and fall of a reindeer herd*, The Scientific Monthly [Science], 73 (6): 356-362. Note that 'reindeer' and 'caribou' are two names for the same species.

Page 133, "And yet, very few of the scientists..." The UN's Food and Agricultural Organization published their estimates of the percentage of people who are undernourished in their 2014 report, *The State of Food Insecurity in the World*.

Page 136, "The obsolete idea that many species..." V.C. Wynne-Edwards put forth his ideas in his 1962 book, *Animal Dispersion in Relation to Social Behaviour*, published by Oliver & Boyd, London.

Page 136, "Wynne-Edwards catalogued..." Wynne-Edwards sings the praises of infanticide and cannibalism as mechanisms of population regulation on page 546 of the book cited above; he cites Carr-Saunders on pages 493-494. The original source was: A. M. Carr-Saunders, 1922, *The Population Problem: A Study in Human Evolution*, Clarendon Press, Oxford UK.

Page 137, "During the ten or twenty years..." Wynne-Edwards was so roundly debunked that his once-considerable reputation faded, and as of the time of this writing, his name has sunk into utter obscurity. He didn't even have a Wikipedia "stub" page until 2005, and his book has been cut of print for decades. All of that is a shame, because he launched an important debate, even if he was wrong.

When I say that by 1990, evolutionary biologists had given up on non-kin group selection, I should mention the most conspicuous exception, who was (and is) David Sloan Wilson. His skeptical colleague, Richard Dawkins, doggedly kept track of Dr. Wilson's efforts to defend his rather awkward position for several years, tirelessly debunking his points one by one. One of their exchanges can be seen in this pair of papers: D. S. Wilson and E. Sober, 1994, *Reintroducing group selection to the human behavioral sciences,* Behavioral and Brain Sciences, 17 (4): 585–614. And: R. Dawkins, 1994, *Burying the vehicle,* Behavioral and Brain Sciences, 17 (4): 616-617.

Page 137, "So what's wrong with that?" Economist Garrett Hardin coined the term "the tragedy of the commons" in 1968, in an essay called simply, *The tragedy of the commons* (Science, 162 [3859]: 1243–1248). The term formally describes the necessary failure of purely cooperation-based systems in which individuals with competing interests share a resource base. Hardin's idea was merely an extension of the 'prisoner's dilemma' problem of game theory to the field of group economics. It is generally believed that the essential problem with cooperation-based systems was first formally identified at RAND in 1950 by Merrill Flood and Melvin Dresher, during (unpublished) efforts to prepare a template for US nuclear strategy in the upcoming Cold War. The "commons" in that case was the un-nuked world we all had to share, and the "tragedy" was that either the US or the USSR could get a real advantage by pushing its button first. The solution to the dilemma was the Mutually Assured Destruction strategy, which both superpowers subsequently pursued. Unfortunately, the MAD strategy doesn't have an analog in most group selection scenarios.

Page 138, "Once famine and the other blights..." Nigeria has 193 people per square kilometer (far above the global average of 49), $5600 per capita GDP (in 2015 US$ at PPP), and a total fertility rate of 5.25 children per woman (far above the break-even level of 2.1). The United States has 33 people per square kilometer, $54,600 per capita GDP, and a total fertility rate of 2.01. TFR and population density data are from the CIA's 2015 *World Factbook*. Per capita GDP estimates are from the World Bank's 2015 *Databank*.

Page 140, "The notion of economic birth dearth..." Schumpeter's comments mocking the idea of birth dearth are found on page 115 of his 1942 book, *Capitalism, Socialism and Democracy* (Harper Perennial, NY).

Page 142, "Consider this fact:..." The statement that one US worker in 2013 generated as much real value as 3.8 workers in 1947 is derived from the following numbers. In 1947, there were 57,038,000 US workers, generating a GDP of $1.76 trillion (adjusted to 2015 value). In 2013, there were 143,929,000 US workers, generating a GDP of $16.8 trillion (2015 value). The GDP data are from Angus Maddison, 2003, *The World Economy: Historical Statistics,* OECD; and *World Development Indicators*, World Bank, 2015. The labor data are from the US Bureau of Labor Statistics.

Regarding the last sentence of this paragraph: Republican president and five-star general Dwight Eisenhower coined the phrase "military-industrial complex" during his televised farewell address on January 17, 1961. He said: "[The] conjunction of an immense military establishment and a large arms industry is new in the American experience... In the councils of government, we must guard against the acquisition of unwarranted influence, whether sought or unsought, by the military-industrial complex. The potential for the disastrous rise of misplaced power exists and will persist."

Page 143, "Almost all statistics involving..." The statement that one US farm worker (plus innovations) could create as much real farm output in 1989 as 4.1 workers in 1948 is derived from the following data. There were 7.63 million documented farm workers in the US in 1948, generating $109.1 billion in farm productivity (expressed in 2005 US$). In 1989, there were 3.2 million documented workers, and an estimated 15% more undocumented ones, for a total of 3.67 million farm workers. They generated $212.5 billion in farm productivity (again in 2005 US$). Data on documented workers is from the US Bureau of Labor Statistics. Data on farm output and undocumented workers is from the US Department of Agriculture.

The argument that agriculture gives good support to the general interpretation of these labor data hinges on the assertion that we can account for undocumented foreign labor on US farms in a reasonable way. The USDA estimates that prior to 1990, illegal workers only accounted for about 15% of U.S. farm labor, though that figure is around 50% today (NAFTA came into force in 1994). My analysis uses the set of assumptions that is least supportive of my conclusions, namely that there were effectively no illegal migrant workers in 1948, and that their presence rose steadily to 15% in 1989. Despite the dampening effect of these assumptions, the growth of per-worker productivity on US farms in this period was amazingly steep, as reported above.

Page 145, "In general, birth dearth..." One interesting aspect of the population decline in the Eastern European and ex-Soviet nations is the role of religion in maintaining birth rates in some nations. This is discussed in Chapter Nine, and in Appendix 2, Topic F.

Page 146, "As in the economic case..." The Battle of Cajamarca is described in: Kim MacQuarrie, 2012, *The Last Days of the Incas*, Hachette, Paris. The Battle of Rorke's Drift is described in: Donald R. Morris, 1965, *The Washing of the Spears*, Simon and Schuster, NY.

Page 151, "The world's first fertility-enhancement..." Iran's Bill 446 was reported in: Amnesty International, 2015, *Iran: You shall procreate: Attacks on women's sexual and reproductive rights in Iran*, Report MDE 13/1111/2015. The 2014 estimate of Iran's fertility rate is from the CIA's 2015 *World Factbook*. Iran is a developing nation, so it might seem odd that the birth rate is so low... but in fact, the situation in Iran is not unusual. The details are discussed in Appendix 2, Topic E.

Page 152, "India has been encouraging..." Statistics regarding India's fertility reduction program are highly contentious. One view, purporting that 4.6 million women were sterilized under possible coercion in 2002-2003, was presented anonymously in the November 7, 2003, New York Times, under the title: *For Sterilization, Target is Women.*

Page 152, "By comparison, the one-child policy..." China's one-child policy is blamed for indirectly causing many parents to selectively abort female fetuses, for reasons of family economics. I personally think that's awful, if it's true, and I refer to the policy as a "success" only in terms of achieving its stated goals, apparently without mass forced sterilization. The estimate of China's 2014 fertility rate is from the CIA's 2015 *World Factbook.*

Page 153, "In 2014, China relaxed its one-child policy..." Roughly two million couples in China were offered licenses to have a second child in 2014, according to the China Mail, July 10, 2014: *Most Chinese provincial areas relax one-child policy.* But fewer than one million actually applied for these licenses, according to the China Mail, January 12, 2015: *1 million Chinese couples apply to have second child.* The survey that showed that the others overwhelmingly claimed that they couldn't afford to raise a second child was reported in the China Mail, November 27, 2014: *Cost of second child puts couples off.*

Page 154, "Banning contraceptives is even more..." The report that 8.6% of US women have experienced efforts to get them pregnant against their will comes from a survey of 9086 women entitled, *The national intimate partner and sexual violence survey*, published in 2010 by the US Centers for Disease Control and Prevention (CDC). The statement that there are *no* laws prohibiting reproductive coercion (e.g., sabotaging birth control) in the US is hard to prove, because it's a negative claim, and the total body of federal, state and local laws in the US is immense. But the statement is commonly made, and I know of no effort to provide any counterexample from jurisprudence, nor any example of a man jailed for impregnating a woman during consensual sex. Wikipedia's article, *Reproductive coercion*, includes the sentence: "Reproductive coercion is not criminalized in the United States" (as of April 2015). The same claim is made in: Public Health Watch, 2013, *Recognizing Reproductive Coercion For What It Is: A Crime*, published online at publichealthwatch.wordpress.com.

Page 158, "That reputation was well justified." The data on relative fertility of US Catholics and non-Catholics in the 1950s and 1960s is from C. F. Westoff and E. F. Jones, 1979, *The end of "Catholic" fertility*, Demography 16(2): 209-217.

Page 158, "But then, in the late 1960s, Catholic fertility in..." The *Humanae Vitae* was published at the peak moment of world awareness of the gravity of the population explosion crisis. Paul Ehrlich's book, *The Population Bomb* (Buccaneer Books, Cutchogue NY) was published almost exactly simultaneously. Pope Paul VI wrote in the *Humanae Vitae* that he was "fully aware" of the population problem in developing countries, but declared: "No one can, without being grossly unfair, make divine Providence responsible for what clearly seems to be the result of misguided governmental policies."

Page 158, "And yet, despite this direct, long-term exercise..." The proportions of the populations of Italy and Poland who are Catholic are 2010 figures, from: *Religion and Public Life*, 2011, Pew Research Center. The recent fertility rates of Italy and Poland are 2014 figures for the TFR (total fertility rate), from the 2015 *CIA World Factbook;* the 1968 fertility rates of Poland and Italy are from the Central Statistical Office of Poland, and Italy's Istituto Nazionale di Statistica. Per capita GDP figures in Poland and Italy are expressed in 2015 US dollars; the 1968 values are adjusted from Angus Maddison, 2003, *The World Economy: Historical Statistics*, OECD, and the 2013 values are from *World Development Indicators*, World Bank, 2015.

Page 158, "Although Catholicism is a conspicuous example..." Several sociological studies in the US and Europe have reported positive correlation between religiousness and fertility, across many religious faiths. For example: T. Frejka and C. F. Westoff, 2006, *Religion, religiousness and fertility in the U.S. and in Europe,* Max Planck Institute for Demographic Research, Working Paper 2006-013. And: S. R. Hayford and S. P. Morgan, 2008, *Religiosity and fertility in the United States: the role of fertility intentions,* Social Forces, 86 (3): 1163–1188.

Page 159, "For example, the Third Reich ran a propaganda campaign..." Estimates of the success of the *Lebensborn* program are from: E. Simonsen, 2005, *Children in danger: dangerous children,* pp 269-285 in *Children of World War II: The Hidden Enemy Legacy,* eds. K Ericsson & E Simonsen, Berg Publishers, Oxford UK.

Page 159, "A government can also try setting up a sort of state religion..." An excellent explanation of *Juche* is found in: G. Lee, 2003, *The political philosophy of* Juche, Stanford Journal of East Asian Affairs, 3: 105-112.

Page 164, "Here's what that tells us." The full passage of the dialogue paraphrased here can be found on page 736 of: Leo Tolstoy, 1877, *Anna Karenina,* Signet Classics, NY. In my paraphrased version, the passage in quotations is verbatim.

Page 164, "The reason I suggest this possibility is that..." If you'd like to see the original literature showing the correlation of child-rearing and unhappiness, a number of the studies are summarized and cited in: N. Powdthavee, 2009, *Think having children will make you happy?* Journal of the British Psychological Association (online), 22: 308-311.

Page 165, "Voluntarily sterile people are already..." The data showing childlessness up from 10% to 18% are from: G. Livingston and D. Cohn, 2010, *Childlessness Up Among All Women; Down Among Women with Advanced Degrees,* Pew Research.

Page 168. The title of this chapter plays off the term 'extinction vortex', which is commonly used in conservation biology. The idea of an extinction vortex is that some population variable, such as genetic diversity, is negatively affected by the falling population of some species, and exacerbates the problem, speeding the species toward extinction. The term was introduced in: M. E. Gilpin and M. E. Soulé. 1986, *Minimum viable populations: processes of species extinction,* Pp. 19–34 in: M. E. Soulé, *Conservation Biology: The Science of Scarcity and Diversity,* Sinauer, Sunderland, MA. An extinction vortex is an example of what is called 'positive feedback' in information theory: a self-regulatory mechanism that enhances initial perturbations rather than correcting them. Such mechanisms tend to make unstable systems even more unstable, hastening their catastrophic collapse. The reason I swiped the term from Gilpin and Soulé, and altered it slightly to fit the human case, is because the mechanism described in this book, which will drive humans extinct, is also a positive feedback system causing systemic instability. But the similarity ends there.

Page 170, "The main engine that drives the model is rising prosperity." Historical estimates of GDP and population have been published by the World Bank, the IMF (International Monetary Fund), the CIA, and the OECD (Organization for Economic Co-operation and Development; for citations, see Appendix One). Of these four, the OECD's study is the most idiosyncratic, and also the boldest: It estimates the per capita GDP (or equivalent) for most of the existing nations over the past several hundred years, and gives annual estimates back to 1950, much farther back than other data sources. There is no doubt that this long reach causes the OECD database to veer further from certainty than its competitors. But, on the other hand, it would be ridiculous to claim that nothing useful can be said of the economies of nations before (say) 1990.

The estimate that global per capita GDP at purchasing power parity is $14,400 (in 2015 US$) comes from *World Development Indicators,* World Bank, 2015.

Page 171, "The model predicts that a grand total of..." Carl Haub estimated that 107.6 billion people had ever been born in his 2011 update of his paper: C. Haub, 2002, *How Many People Have Ever Lived on Earth?*, Population Today, 30: 3–4. In the time since that update (from mid 2011 to late 2015), about 600 million more people have been born, bringing us up to a bit over 108 billion overall.

Page 174, "Another subject that hasn't come up..." The concept of the technological singularity was attributed to von Neumann by Stanislaw Ulam in his "Tribute to John von Neumann", published in 1958 (Bulletin of the American Mathematical Society 64 [3 pt 2]: 1-49).

Hawking made his remarks in an interview with the BBC on December 2, 2014, reported under the title, "Stephen Hawking warns artificial intelligence could end mankind." At the time of this writing, the article is available online at http://www.bbc.com/news/technology-30290540. Musk called AI "our biggest existential threat" while addressing MIT students at the AeroAstro Centennial Symposium, as reported in: *Elon Musk: artificial intelligence is our biggest existential threat*, Guardian, 27 October 2014.

Page 175, "In Chapter Eight, we explored the role of..." The idea that we are living in the second phase of automated productivity, a historical period in which simple, non-creative mental labors are replaced by computers, is explored in depth in the 2014 book, *The Second Machine Age*, by Erik Brynjolfsson and Andrew McAfee, published by W. W. Norton, NY.

Page 175, "Most commentators who have expressed worries..." Hawking and Musk were among the most prominent of the 20,000 signatories on a 2015 petition to ban armed AI systems. See: *Musk, Wozniak and Hawking urge ban on warfare AI and autonomous weapons*, Guardian, 27 July 2015.

Page 182, "There are two major lacunae..." The OECD reports that, during the years from 1950 to 2000, most of the oil-rich countries of the Middle East were much less wealthy than other international observers believe (IMF, World Bank, CIA). Because of this discrepancy, I have removed Qatar, Libya, Iraq, Bahrain, Kuwait, Oman, the UAE and Saudi Arabia from the analysis completely. The other OECD estimates of GDP regress very closely with the other sources (at least as closely as they regress to each other), but at about 75% of the absolute values estimated by the other three. This requires no re-adjustments in terms of trends and correlations, but I have multiplied the OECD's estimates of GDP by an adjustment factor of 1.33 when comparing them directly with other sources. None of these adjustments should be construed as suggesting that I think the OECD's estimates are less reliable than other sources; I have merely followed the most parsimonious protocol to make all the sources commensurate.

Page 190, "Note that the United Nations..." The UN bases its projections upon estimates of the volatility of the first and second derivatives of population size. In other words, they acknowledge that, in any given region of the world, population growth rates are changing, and that the rate of that change is also changing. They furthermore acknowledge that those changes are not consistent, but fluctuate over time. They attempt to characterize the fluctuations by a probabilistic volatility function, and then to give a range of future scenarios in which the population goes up or down. A lot of this seems to be done pretty informally, but I am not privy to their exact methods. This general way of doing things was originally developed for analyzing stock markets, and assigns probabilities to future outcomes based upon various assumptions about the nature of (and causes of) volatility. As in the case of financial analysis, there's no way to ever know if those assumptions were right or wrong... only if a given prediction fails or succeeds. If an analyst is wrong, he or she can always say, "Well, boss, I told you there was a 10% chance this was going to happen." If the chance was actually 70% back when the prediction was first made, well... who can prove it now?

My model uses an evolutionary observation about the relationship between economic growth and the intrinsic human drive toward procreation. UN modelers have never taken this factor into account. But, even *their* methods acknowledge a 20% chance that future global population will follow a trajectory at least as low as the one I predict. I hope the arguments given in this book are sufficient to explain why I feel that a trajectory in that range is, in fact, a certainty.

INDEX

www.ingramcontent.com/pod-product-compliance
Lightning Source LLC
Chambersburg PA
CBHW051727260326
41914CB00031B/1781/J